INGREDIENT

INGREDIENT

UNVEILING THE ESSENTIAL
ELEMENTS OF FOOD

ALI BOUZARI

ecco

An Imprint of HarperCollinsPublishers

HarperCollins books may be purchased for educational, business, or sales promotional use. For information please e-mail the Special Markets Department at SPsales@harpercollins.com.

FIRST EDITION

Photographs by Jason Jaacks
Illustrations by Jeff Delierre
Designed by Suet Yee Chong

Library of Congress Cataloging-in-Publication Data has been applied for.

ISBN 978-0-06-238535-2

16 17 18 19 20   OV/RRD   10 9 8 7 6 5 4 3 2 1

cover more ground (see the Heat chapter). With their newfound agility, hot water molecules can run circles around their captives. Fewer water molecules are required to keep each Ingredient in check, so more can dissolve in the same amount of water. We can use heat in the kitchen to pack a lot of Ingredients into small amounts of water, helping us create concentrated syrups, sauces, stocks, brines, extracts, and drinks like coffee and tea. The exception to this rule is gases. As the temperature goes up, gases become even nimbler than the water molecules, slipping away from the watery mob and escaping into the air. This is why we keep soda, beer, and champagne as cold as possible: it helps the gases stay dissolved, preserving fizziness longer.

When we steep flavorful bits of food in water, we're pulling out a certain mixture of the various Ingredients that were locked in that food. The exact ratio of Ingredients that come out of the food depends on how well each Ingredient dissolves in water at that temperature, so varying the amount of heat allows us to create different combinations of taste, aroma, color, and texture, even when using the same starting material. Most of us are familiar with the different flavor profiles that come from cold-brewed versus hot-brewed coffee, and we can apply the same approach to anything that involves dissolving delicious things in water (this also applies to dissolving water-hating stuff in lipids—see the Lipids chapter). We can steep tea, mull wine, and even make stock at temperatures ranging from freezing to boiling to yield countless variations on a central flavor theme.

texture, taste, aroma, and appearance of your food. It'll be like cooking with X-ray vision.

The Ingredient patterns show us that there are only a couple of basic options for fixing any culinary problem. Here's an example: When a recipe for a dumpling, sausage, or cookie fails (it crumbles to pieces, for instance), you'll know that the two Ingredients with the greatest power to stick stuff together are proteins and carbs. In your head, you'll picture long strands of these Ingredients lashing together to create webby nets that help food hold its shape. You'll know where to find them—root vegetables, meat, fruit, or a bag of flour, to name a few. You'll also know how to treat those ingredients in order to unpack the long, tangled Ingredients within—mashing the root vegetables aggressively, grinding the meat finely, simmering the fruit rinds for a long time, or dispersing the flour properly. All of these are variations on the same basic principle, based on personality quirks of carbs and proteins, and you'll be able to choose the one that best suits your tastes.

These patterns apply to more than just failures: If something is good, you'll know what made it good and how to do it again. If you've made something a thousand times and you want #1,001 to be even better, you'll see how. You'll know that crispiness comes from the balance of water to the other Ingredients, so that's the best place to start your quest for an even crispier crust on your famous pizza. If you want to substitute ingredients in a recipe because of allergy, aversion, diet, or being too lazy to go to the store, you'll know what your options are. When a gluten-free friend comes over for dinner, you'll know that you can replace the thickening

power of wheat flour in your beef stew with any mixture of carbs and proteins—everything from okra and pureed parsnips to ground hazelnuts and crumbled corn tortillas. If you want to explain a recipe to a friend, family member, or employee, you'll have a language that he or she can understand. Your uncle will have a better chance of success with your roasted carrot recipe if you mention that the butter is in there to contribute proteins and sugars for browning, not just because you put butter on everything. If you have a crazy idea for something that no one has ever made before, you'll save a lot of time getting there. A melon ice cream that you can shave into tissue-thin sheets becomes a more attainable idea when you know that carbs and proteins will provide structural support to hold the sheets together, sugar will keep water crystals small and the ice cream smooth, lipids will carry the aroma of the melon, and removing gas bubbles from the ice cream will make it dense enough to shave.

There's a lot to learn in this book, but it's presented as getting to know eight characters in a story rather than memorizing a bunch of facts and figures. Each Ingredient stars in its own chapter, where its key personality traits shine under the spotlight. Everything is written in 100 percent humanspeak, no scientific background required. To make the personality of each Ingredient come to life and stick in your mind, I've included beautiful illustrations of all the metaphors. Courtesy of the talented and innovative artist Jeff Delierre. Jeff rendered each chapter in a different medium to match each Ingredient's personality: watercolor for the water chapter, oil paint for the lipids chapter, etc. After you've

made it through each concept, your reward will be a two-page photo spread shot by Jason Jaacks, who is one of my best friends, a National Geographic Explorer, and the most engaging visual storyteller I know. The photos group strange combinations of foods together—sea urchins nestled next to French fries and a purple artichoke, for example—to hammer home the universality of each concept and show how these simple patterns explain everything we cook and eat.

Ingredient works well as a reference book, but it's meant to be treated like a mini curriculum, a crash course in how food works. Read it beginning to end and you will gain a Sherlock Holmes–like superpower to help you visualize the patterns, connections, and solutions to questions you come across in the kitchen for the rest of your life.

WAT

watercolor

ER

Water is important: It's the theater in which the other Ingredients perform. It changes the way the other Ingredients act. Water is the key to understanding most of this book, and luckily, water follows five simple, universal rules in food:

- It is **solid, liquid,** or **gas**.
- It **dissolves** things.
- It **flows**.
- It is **acidic, neutral,** or **basic**.
- It makes things **grow**.

SOLID, LIQUID, OR GAS

Water can transform from solid to liquid to gas. There's nothing special about that—anything will change to those forms if the temperature is right. Solid salt, for example, can liquefy and evaporate but only at temperatures hot enough to literally melt your face off. Water is special because it can change between all three forms at temperatures closer to your mouth's comfort zone. Controlling water as it transitions from solid to liquid to gas and back again is just as important for baking soufflés as it is for microwaving Hot Pockets.

liquid

Food is mostly liquid water. Zoom in far enough, and most food looks like a sea of tiny droplets. That sea is littered with chunks, blobs, strings, and bubbles of stuff. Those chunks, blobs, strings, and bubbles are the other Ingredients, such as sugars, lipids, carbs, minerals, gases, and proteins. The swirling sea of water gives the other Ingredients room to drift around, and heat gives them the energy to do so. That movement sets the stage for nearly everything that happens in cooking.

Liquid water calls the shots in liquid *and* solid food. Milk, honey, and broth obey the same "water rules" as raspberries, carrots, and chicken wings. Unless they're completely frozen or bone-dry, solid foods only *look* solid. Solid food is mostly liquid water, but Ingredients like proteins, lipids, and carbs form walls that

corral the water into a maze of chambers. Without these walls to hold liquid water in place, a piece of celery would be a shapeless puddle.

> The juiciest food has chambers that are full of water *and* ready to burst.

When you bite into a steak, an apple, or a piece of fresh mozzarella, you tear the tiny water chambers open, releasing a flood of juice. The juiciest food has chambers that are full of water *and* ready to burst. Raw steak and underripe peaches have chambers full of juice, but the walls around those chambers are too strong to give up the goods. Cooking and ripening weaken the walls to the point where the slightest pressure causes them to explode, delivering the gushing revelations of medium-rare steak and peaches at peak ripeness. If the steak is overcooked or the peaches sit on a shelf for too long, the chambers dry out, and no amount of chewing will release any juice.

ice

When liquid water freezes, the randomly flowing sea of droplets transforms into a rigid, immobile iceberg. That transformation is all about organization. First, a couple of water molecules congregate around an anchor point—any impurity that they can hold on to, like a speck of dust, a bubble, or an imperfection on the side of a glass. Around this anchor point, the water molecules form a tiny ice crystal. Over time, more

water molecules flock to the crystal, snapping into perfectly straight geometric shapes as it grows outward.

When water freezes, it makes food firmer. We can use that firmness to hold any floppy, runny, or squishy food in place while we slice, grind, or shape it to our liking. Freezing helps us slice meat into paper-thin carpaccio without it mushing apart, scrape snowy granita off of a chunk of frozen juice without it turning into sludge, pulverize leathery chipotle chilies into a powder without it clumping, cut a fragile cake into perfectly clean slices without them crumbling, or form a delicate soup dumpling around frozen broth without it leaking everywhere.

Things like frozen fish sticks and mango chunks may feel like solid blocks of ice, but all frozen food is actually a mixture of ice crystals and liquid water. Pure water is the only thing in the kitchen that can freeze into one big mass of ice crystals, regardless of how burly your freezer is. Freezing requires organizing water into perfect rows and columns, and that organization is difficult when other Ingredients are littered around. Everything we eat is full of sugars, proteins, carbs, lipids, gases, and minerals that get in water's way as it tries to form pure crystals. Instead of a giant block of ice, we get thousands of small water crystals surrounded by a concentrated syrup. The syrup contains the water that was too tangled in the other Ingredients to make the journey to join a crystal. This happens naturally in all food, but we enhance the effect in ice cream, sorbets, and frozen margaritas by adding extra Ingredients to give a range of textures from slushy and grainy to smooth and creamy.

In liquid form, water molecules are free to move, so they can mush together like people on an extremely crowded dance

Instead of a giant block of ice, we get thousands of small water crystals surrounded by a concentrated syrup.

floor. In an ice crystal, the rigid spacing of the rows and columns holds each molecule at arm's length from its neighbor. This makes water expand when frozen, and it can cause damage. Anyone who has ever left a beer in the freezer for too long knows what can happen when water expands in a tight space. Now imagine thousands of crystals growing inside every piece of food. Expanding crystals act like icebergs breaching the hull of the *Titanic,* tearing and shredding food from the inside. When that food is thawed, those crystals shrink and retract, opening holes in the watery chambers and releasing a flood of juice. This can make shrimp soggy and mealy, which is bad, or it can help you get more juice out of a blueberry, which is good.

The balance between frozen and unfrozen water dictates the texture and quality of frozen food. Heat provides us with another way to control that balance (see the Heat chapter). Adding heat gives water molecules the energy to flop around more restlessly, preventing them from standing still long enough to hold together as frozen ice. When ice melts, the crystals shrink as molecules ooze out of their meticulously crafted ranks to resume life in liquid form.

Removing heat obviously helps food freeze, but the speed at which we remove heat and the total amount of heat that we take out can make a huge difference in the texture and shelf life of our food. Bringing the temperature down quickly gives the ice crystals less time to grow before everything slows down due to lack of energy (see the Heat chapter), and smaller crystals can fit in the spaces between cells more comfortably. These smaller ice crystals are less likely to pierce and destroy, which is why a lot of expensive things like truffles, lobster, and sashimi-grade tuna are quick-frozen to preserve quality. This isn't a solution that works indefinitely, however, since even small ice crystals will continue to grow slowly as stray, unfrozen liquid water molecules travel to join them over days to weeks in a

freezer. To truly keep ice crystals from growing during storage, we need to remove enough heat to get everything to stop moving altogether. When food gets cold enough, the thick liquid trapped between the ice crystals will solidify without crystallizing. Food locked in this "glassy" state (see also the Flow section) is about as deeply frozen as it can be—things can be held this way almost indefinitely with very little change. While the freezing equipment to get food down into the −40°C range (where most foods become glassy) and below is expensive, using it to preserve a tuna loin bought for the price of a car might help rationalize the expense.

steam

In steam, everything is so spread out that most water molecules never even touch each other. If liquid water is a crowded dance floor, steam is outer space. We only actually eat a minuscule amount of steam—it usually escapes before we get the food into our mouths—but steam plays a key role in everything from croissants to pork rinds.

When liquid water turns into steam, the molecules take flight, blasting off into the air in random directions. They spread out to take up more than a thousand times the space of liquid water. The airborne water molecules slam into anything in their way, exploding outward as they escape. One or two water molecules flying around won't affect anything on a human scale, but millions of water molecules vaporizing together generate force like a tiny volcano. Creating that much pressure inside the hull of a popcorn kernel, the skin of a grilled vegetable, the breaded crust of a fried chicken leg, or the surface of a soufflé causes

those things to pop, tear, puff, and rise. Water is the fuel that drives these explosive changes, and balance is key to get the texture we want. Food with too little water lacks the force necessary to erupt, and food with too much water stays soggy after cooking. Whether the food is baked, fried, or grilled, adjusting the water in a recipe can be the difference between light and crunchy, and dense and chewy.

It takes a lot of energy to send water into orbit, and flying water molecules take heat with them as they leave the food. Those escaping molecules keep the temperature of the food from rising above the boiling point of water. Under normal conditions, water puts a ceiling on how hot food can get. When we remove water, the temperature can rise uninterrupted, and food can quickly turn golden brown. Everything from bread and mushrooms to French fries and steak starts to brown quickest once the surface dries off. Dabbing the water off of scallops with a towel before searing is an old chef trick to get a head start on surface drying, opening the earliest possible window for the scallop to brown in the pan before overcooking.

Apart from removing water from the food's surface, the only way to get the temperature above the boiling point of water is with pressure. Pressure cookers trap water molecules as they try to rocket out into space. Since the water can't leave, it stays put, bringing the temperature of the food above the boiling point even in the interior of the food. This makes pressure cookers one of the fastest forms of cooking we've invented so far.

> Under normal conditions, water puts a ceiling on how hot food can get.

Water molecules form a rigid grid in ice, flop around
in water, and zip through space in steam.

CROISSANT

LETTUCE

FROZEN
SALMON

WATER / SOLID, LIQUID, OR GAS

POPCORN

Water expands as steam, puffing and leavening.

STEAK

When bitten, liquid water is released and makes things juicy.

ICE POPS

Solid water crystals make things firm.

DISSOLVING

Stir some sugar into a glass of water, and the crystals fade away. The process seems gentle, as if the sugar peacefully drifts off into oblivion. It doesn't. Water descends on each crystal like a rabid mob, ripping it apart. **Each individual sugar molecule gets torn out of the crystal and carried away by a pack of water molecules.** Water can do this to nearly every Ingredient—it draws carbs out of fruits and vegetables as they simmer in a sauce, dismantles proteins in braised meat, extracts sugars and minerals from tea leaves, and traps gases in carbonated drinks. Lipids, however, are defined by the fact that they generally hate water. Letting water loose on the other Ingredients helps us flavor, preserve, and change the texture of food.

Dissolving is a two-way street—water holds the Ingredient captive, and the Ingredient holds water's undivided attention. Both water and the Ingredient are held together, trapped in a force field "cocoon." This isn't a traditional couple's dance; each Ingredient molecule gets an entire entourage of water. Thousands of water molecules can glob on to a single Ingredient. This means that, at some point, all of the water molecules become spoken for, and the water becomes saturated.

In a saturated sugar syrup, no more sugar will dissolve, unless we add more water or heat. **Adding heat helps the water molecules zip around faster, so they**

> Dissolving is a two-way street—water holds the Ingredient captive, and the Ingredient holds water's undivided attention.

These people asked very different questions, but the answers I gave them always touched on the same thing: the difference between ingredients and *Ingredients*.

Potatoes, cheese, caviar, duck fat, and pasta are all ingredients. Every ingredient, no matter how complex, is made of something simpler: Ingredients, the basic building blocks of food. There are eight Ingredients: water, proteins, carbohydrates, minerals, gases, sugars, lipids, and heat. The first seven Ingredients are the gears that turn inside everything we eat, and heat is the energy that moves them. Each Ingredient has a personality, a set of things it does or doesn't do, an MO. I knew those personalities—they exposed the basic patterns of how food works, which I used like a Rosetta stone—to answer other people's questions about cooking. This book will teach you those personalities so you can answer your own questions.

This book is not a set of equations you must memorize, nor is it meant to be a definitive treatise on the science of the kitchen. Tons of amazing books, papers, and articles have already been written on the precise physics, chemistry, and biology of cooking. Think of this as the comic book version of those—it'll use metaphors, illustrations, and photos to show you the fundamental principles of how food works. It'll put voices in your head and make you see things . . . in a good way. When cooking and eating, you experience stuff on a human scale—you see, smell, taste, touch, and hear. After reading this book, the voices and images in your head will help you envision the invisible—to picture the microscopic drama behind the

INTRODUCTION

In my early twenties, I got a job teaching at the Culinary Institute of America. There were a lot of terrified students in that building. Some of them had never turned on a stove before. Now they had to learn to cook six different things at once. Perfectly. Consistently. They needed to learn quickly.

At the same time, I stalked my favorite chefs. I developed consulting relationships with some of the best chefs in the world: people like Thomas Keller, Daniel Humm, Corey Lee, and Christopher Kostow. They were *not* terrified. They could cook anything in existence. They wanted to create things that hadn't existed before, to cook food that was brand-new.

I also talked about food with my friends, my little sister, my mom, my grandmother, and random people next to me on planes. Some had questions about recipes they'd read in a magazine, seen on TV, or gotten from a friend. Some wanted to break free from the shackles of recipes altogether, to cook from the wrist with whatever they had in the fridge.

Everyone, from master chefs to my mom, was asking questions about cooking. Sometimes the questions were fancy: "When I wrap paper-thin sheets of potato around a core of molten fondue in the shape of a perfect sphere, fry it in duck fat, and slather the whole thing with caviar, how do I keep the potatoes shatteringly crisp?" Sometimes the questions were not fancy: "Why'd my spaghetti stick together?"

INGREDIENT

CONTENTS

FOR MY DAD

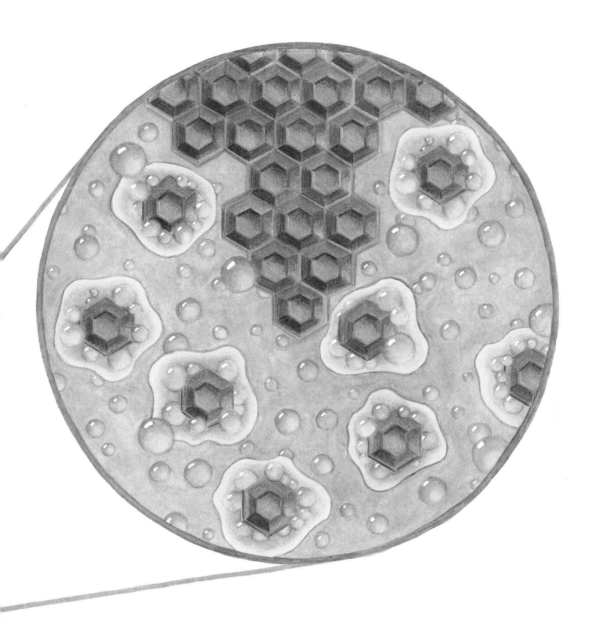

WATER / DISSOLVING

Water dissolves other Ingredients by mobbing each
Ingredient molecule. Once dissolved, both water and
the other Ingredient are bound together.

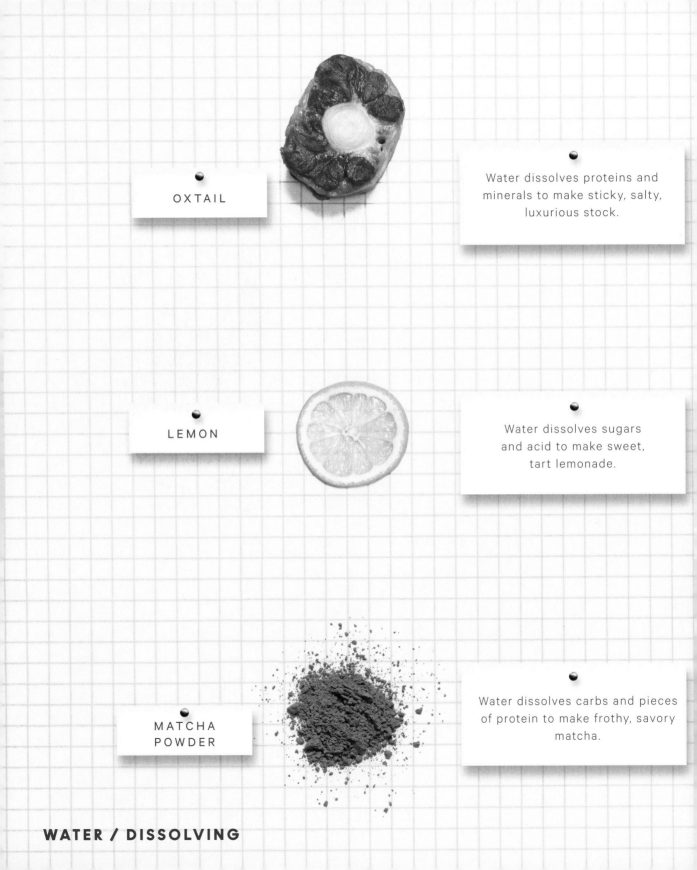

OXTAIL

Water dissolves proteins and minerals to make sticky, salty, luxurious stock.

LEMON

Water dissolves sugars and acid to make sweet, tart lemonade.

MATCHA
POWDER

Water dissolves carbs and pieces of protein to make frothy, savory matcha.

WATER / DISSOLVING

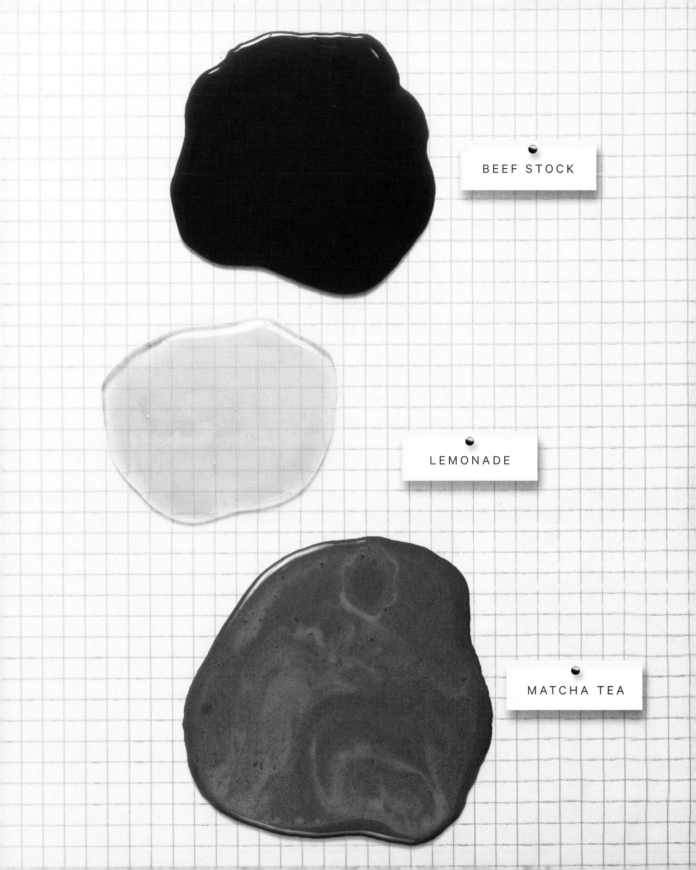

BEEF STOCK

LEMONADE

MATCHA TEA

FLOW

When you shake, swirl, or pour any liquid, the Ingredients in that liquid tumble around. Small Ingredients roll back and forth easily, so they make thinner liquids. Big Ingredients flop and stumble, making liquids thicker. Water is tiny and nimble compared to the other Ingredients, so pure water is one of the thinnest liquids in the kitchen. Adding other Ingredients to water puts hurdles and barriers in its path. Compared to pure water, hummus is an obstacle course.

Any Ingredient roadblock in water's way will thicken a liquid. Scattering obstacles evenly throughout the liquid will thicken it more. Proteins make an egg sauce thicker when they are evenly dispersed than when they clump together from overcooking. Well-whisked, creamy vinaigrette becomes thinner when the lipids separate out to make an oil slick. Carbs thicken lemon marmalade when they are unpacked from the fruit during long simmering. The froth on an espresso is thickest when the bubbles are tiny, and sugary syrups are stickiest when all of the sugar is dissolved. Minerals like salt can also make liquids thicker, but it would take a disgusting amount of salt to have a noticeable effect.

If pure water is among the thinnest liquids in the kitchen, what happens at the other end of the spectrum? As the balance between water and other Ingredients shifts (or we remove enough heat—see the Solid, Liquid, Gas section), food gets thicker . . . and thicker . . . until *everything stops*. The molecules get

> Any Ingredient roadblock in water's way will thicken a liquid.

so jammed up that nothing can move, including water. This kind of substance is called a glass. It's called a glass because it acts like glass—brittle and shatter-y. Potato chips, freeze-dried fruit, baklava, hard candy, Peking duck, bread crust, fried shallots—all of them are glassy foods, and we love them because they're crispy. When you bite into something crispy, the molecules can't slide out of the way of your teeth, so they resist until the breaking point. At the breaking point, the whole structure fails, and the food splinters apart in an explosion of crispiness. **Water is the enemy of crispy—it allows the other Ingredients to slide around, so the food bends instead of breaks.** When we dredge chicken wings, sear steaks, griddle waffles, bake pizzas, fry potatoes, and boil candy syrups, we are removing water and/or adding other Ingredients. Every cooking technique for making stuff crispy is designed to skew that balance toward less water and more of the other Ingredients. Anything that allows water back into the food is a step in the other direction, which is why crispy things get soggy if they're left open in humid air.

> The molecules get so jammed up that nothing can move, including water.

WATER / FLOW

Any Ingredient that gets in water's way will thicken
a liquid. Put enough things in water's way, and it can
stop moving completely, forming a glass.

PURE WATER

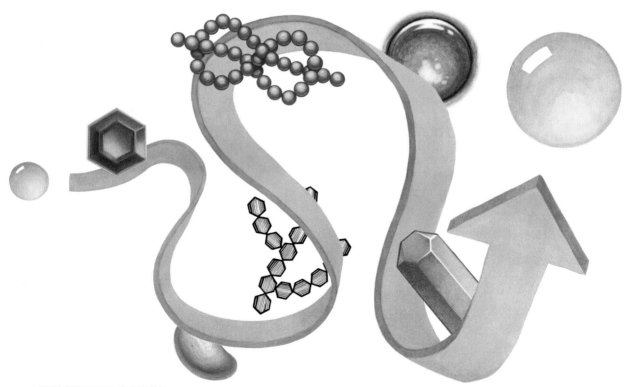

THICKENED SAUCE

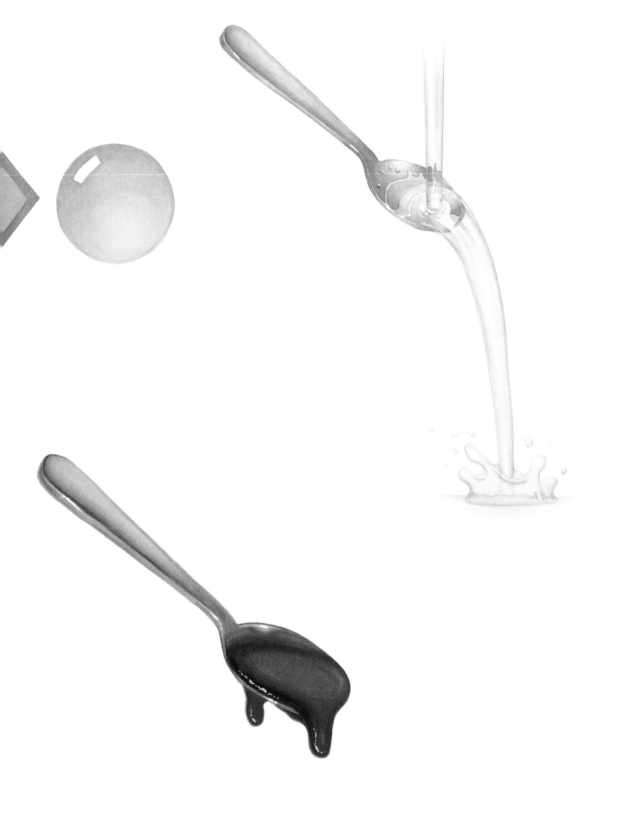

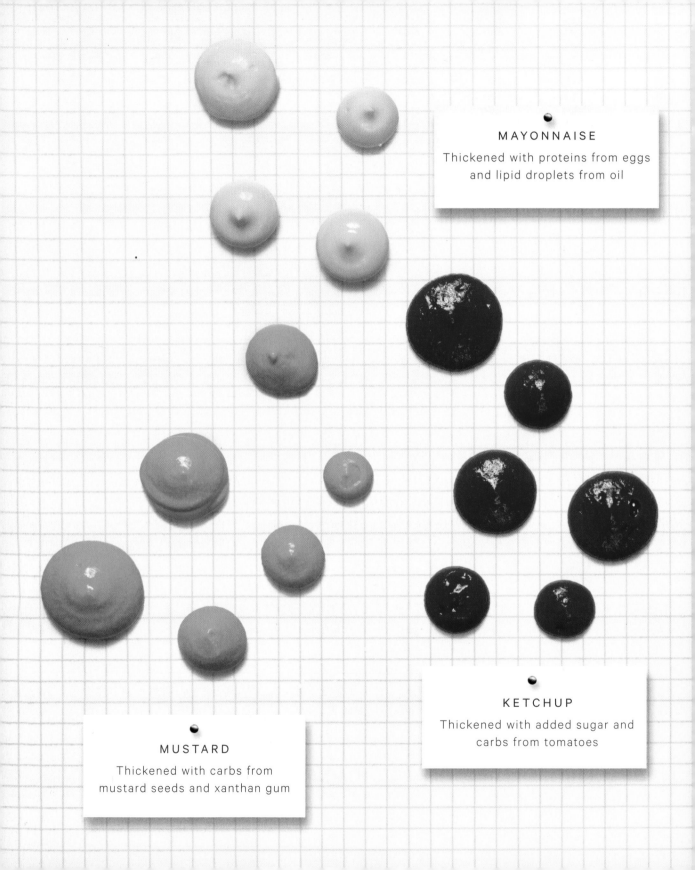

MAYONNAISE
Thickened with proteins from eggs
and lipid droplets from oil

KETCHUP
Thickened with added sugar and
carbs from tomatoes

MUSTARD
Thickened with carbs from
mustard seeds and xanthan gum

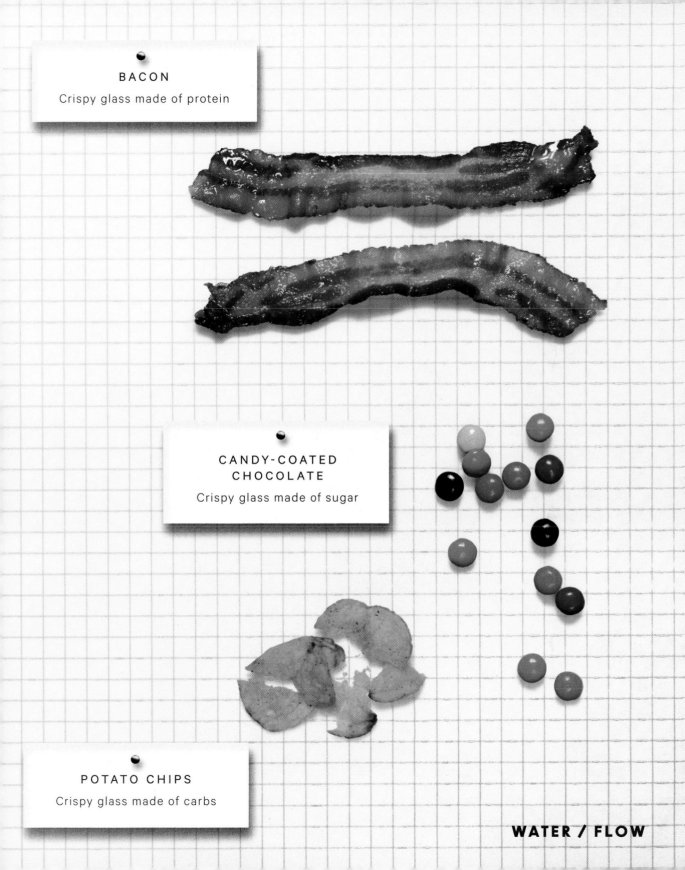

BACON

Crispy glass made of protein

CANDY-COATED CHOCOLATE

Crispy glass made of sugar

POTATO CHIPS

Crispy glass made of carbs

WATER / FLOW

ACIDS + BASES

Water molecules are made of two parts, one acid and one base. Food contains millions of water molecules and thus massive numbers of these halves. In pure water, equal amounts of both halves come together and cancel each other out, making the water perfectly neutral. In most foods, the balance of acidic and basic halves is skewed one way or another. When there are more acidic parts, you have acidic food; with more basic parts, the food becomes basic or alkaline. We measure the balance between acidic and basic parts on a spectrum called the pH scale, which goes from 0 to 14. Neutral water is a 7 on the scale; going down from 7 means things are getting more acidic, and going up from 7 means they are getting more basic/alkaline. Anything that we add to food to lower the pH is an acid, anything that raises the pH is a base.

Food tastes different across the pH spectrum. Acids taste sour, and bases taste soapy or somewhat bitter. Each type of acid has its own characteristic nuance, but creamy lactic acid in butter, bracing acetic acid in vinegar, and fruity malic acid in cherries all share a general sourness. It's harder to describe the flavor of bases because we eat so few foods that have a pH higher than 7—old egg whites and some Dutch cocoa powders are among the exceptions. The slippery bitterness of high-pH foods feels similar to lots of naturally occurring poisons, so those tastes usually make us panic rather than invite us to linger and ponder their flavorful nuances.

Acids and bases affect the structure of the other Ingredients. At either end of the pH spectrum, the structures of the other Ingredients change. Extreme pH swings can make proteins stick together and form webby, structure-building nets, giving firmness to acidic crème fraîche and ceviche as well as alkaline ramen noodles and century eggs. Acids and bases can make carbohydrates disintegrate, which is why baking soda (base) and vinegar (acid) have such drastic effects on the texture of everything from pickled onions to blanched broccoli. They also affect how sugars fuse with proteins to explode during Maillard browning—high pH makes them more fragile and explosive, so pretzel dough is traditionally dipped in an alkaline brine before being baked to give pretzels their characteristic, deep brown color. This high-pH browning trick works on more than just pretzels—root vegetables, bacon, cheese, and anything else that contains sugar and proteins will brown quicker when we make them more alkaline. Slight changes in pH affect the way lipids decay and become smelly, and drastic pH changes can turn them into soap. Altering pH affects how minerals dissolve, which can affect the colors they create in red meat and green vegetables. When certain acids and bases neutralize each other, they generate gas, which is how things like baking soda and vinegar leaven everything from muffins to science fair volcanoes.

Water molecules are made of two parts, one acid and one base.

Pure water is made of two parts: one acid and one base

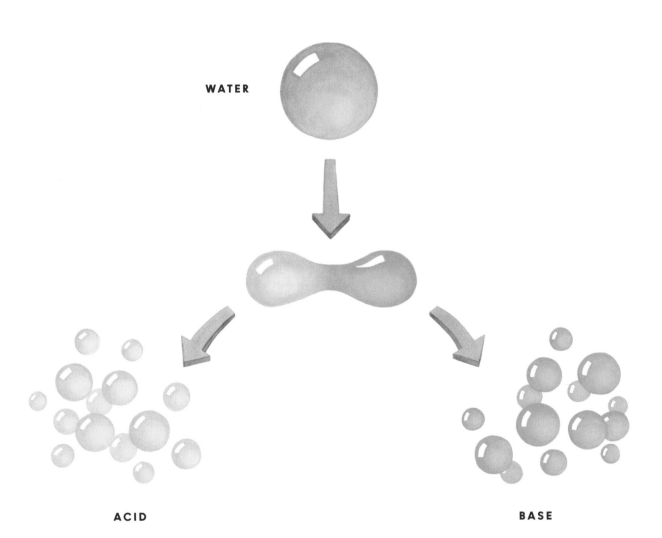

WATER

ACID

BASE

MILK

SOUR CREAM

When the balance between the two parts is skewed, we get acidic or basic foods, which can change their taste, texture, aroma, and color by affecting the other Ingredients.

PURPLE CABBAGE

Neutral—turns red when cooked in acidic water, blue when cooked in alkaline water.

STRAWBERRIES

Acidic—most fruits are.

RAMEN NOODLES

Alkaline—high pH toughens gluten proteins to help noodles stay chewy in hot broth.

WATER / ACIDS + BASES

BLUEBERRY SCONE

Neutral—acid + alkaline reaction creates leavening; too much alkaline (baking soda) makes blueberries turn green.

LAMB LOIN CHOP

Slightly acidic—meat becomes acidic during aging.

CRÈME FRAÎCHE

Acidic—acid causes milk proteins to coagulate and thicken cream.

LIME

Strongly acidic—most citrus fruits are.

PRETZEL

Alkaline—high pH makes browning happen faster.

EGG

Alkaline—and it gets more alkaline with age.

GROWTH

Microbes, bacteria, yeasts, molds, and other living things too small to see are on everything, everywhere, all the time. That is a scary thought. Luckily, microbes need water to survive—and we have lots of ways of taking it away from them. Apart from slaughtering microbes with heat (see the Heat chapter), food preservation is all about water.

Our hairy ancestors figured out that drying food is a great way to keep it from going bad. The explanation is simple: Dried food has less water. When we take water away, thirsty microbes lose their ability to grow. Microbes on the surface of anything that has been sufficiently dried—peaches, jerky, mushrooms, shrimp, rosemary, or lentils—look like withered flowers drooping over cracked, parched earth.

Removing water from food isn't the only way to keep it away from microbes. Freezing and drying have similar effects on microbes even though freezing leaves the water in the food. Microbes grow on food surfaces, slurping and siphoning liquid water from the interior. They can't grow on things like frozen peas and ice pops because the water is locked into a solid grid, immobile and undrinkable.

Water in candied or salted food, while still liquid, is too busy mobbing the dissolved sugars and minerals to pay microbes any attention. Binding water with minerals and

> Microbes need water to survive—and we have lots of ways of taking it away from them.

sugars is how preserves are preserved, why sailors invented salted fish and capers, why we could safely eat the same honey that Cleopatra drizzled on her breakfast, and why candied orange peel doesn't turn into a fuzzy, Technicolor explosion of mold.

If you can't beat 'em, poison 'em. Microbes like neutral water, and changing the pH—making foods acidic or alkaline—dismantles the proteins that microbes need to move, eat, and reproduce. Extreme pH swings kill microbes by short-circuiting their machinery. Since really high pH often tastes terrible, we usually go with the lower end of the spectrum and add acid. The acid can come from two places: from acidic ingredients like fruit juice and vinegar or from helpful, acid-producing microbes that we recruit during fermentation. This chemical-warfare strategy helps acidic pickles, yogurt, kimchi, ceviche, and charcuterie last a lot longer than their neutral counterparts.

MICROBE

RAW APPLE

FROZEN

WATER / GROWTH

Microbes need water, which is plentiful in most raw food, to grow. Freezing, candying/salting, drying, and changing the pH keep microbes from growing, which preserves the food.

CANDIED

DRIED

PICKLED

DRIED PASTA
———
DRIED OREGANO
Little water left

STRAWBERRY JAM
———
DATES
Plenty of water, but it's spoken
for by lots of sugar

PICKLED MUSTARD SEED
———
PICKLED PEPPERS
Water too acidic for microbes
to tolerate

WATER / GROWTH

SALT COD

———

BEEF JERKY

Partially dried, remaining water
bound by minerals

FROZEN GREEN BEANS

Water solid, locked in place,
and undrinkable

colored pencil

SUG

ARS

Sugars are technically a type of carb (see the Carbohydrates chapter), but they are so important to cooking and behave so differently from other carbs that they get their own Ingredient chapter. Say "sugar," and most people think ice cream, honey, candy, pastries, and fruit. Those associations stem from sugar's most famous attribute: sweetness. Sugars do a lot more than just taste sweet—they're as important for good fried chicken and sauerkraut as they are for good cinnamon rolls and maple syrup. Including sweetness, there are six total things sugars do for us in the kitchen:

- They taste **SWEET**.
- They **BROWN** food.
- They **CRYSTALLIZE**.
- They **DISSOLVE**.
- They **THICKEN** liquids.
- They will **FERMENT**.

SWEETNESS

There are dozens of different sugars found in food, but a handful of them are more abundant than the others: glucose, fructose, sucrose, maltose, and lactose. They're all sweet, but each has a different level of sweetness. If we were to rank the sugars from least sweet to sweetest, it would look like this:

lactose < maltose < glucose < sucrose < fructose

The differences in sweetness come from the way each sugar interacts with our taste buds, which work like tiny hands on the tongue. Taste buds experience the world through touch—they grab a sugar, feel its shape, and send a message to the brain. The sweetness enclosed in that message depends in part on the different shapes of each sugar. Sugars like fructose and sucrose send louder messages to the brain than the others.

This opens up some fantastic possibilities. If you don't want sweetness but *do* want to take advantage of the other five things that sugars do for your food, you can use a sugar like maltose. If you want to maximize sweetness with a minimum amount of sugar, using fructose will give you more sweetness bang for your buck. Just ask every soda company.

STRAWBERRY   ICE CREAM   HONEY

TASTE BUDS

Our taste buds grab on to sugars in food and relay sweetness
information, which differs between types of sugar, to the brain.

SUGARS / SWEETNESS

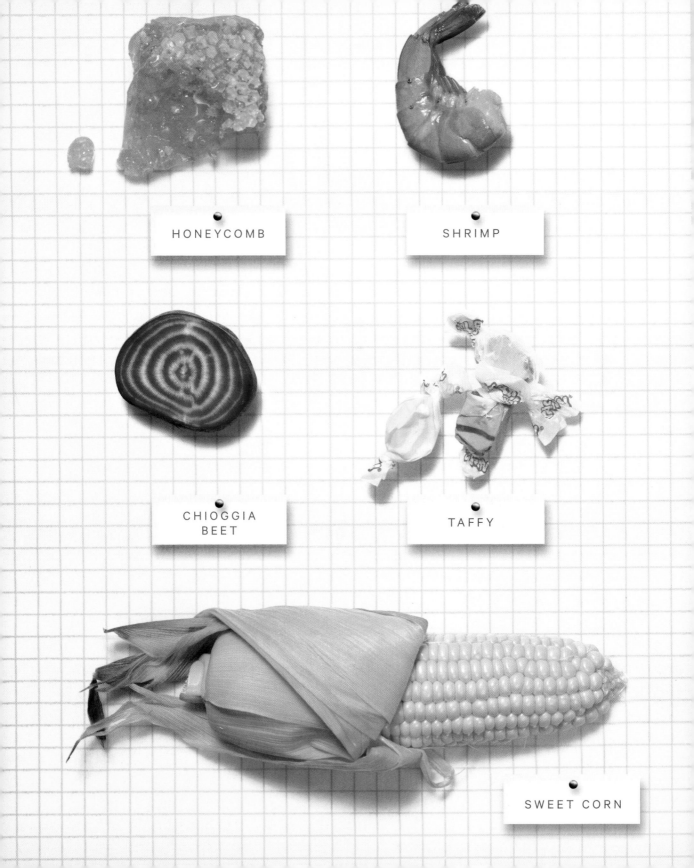

HONEYCOMB

SHRIMP

CHIOGGIA
BEET

TAFFY

SWEET CORN

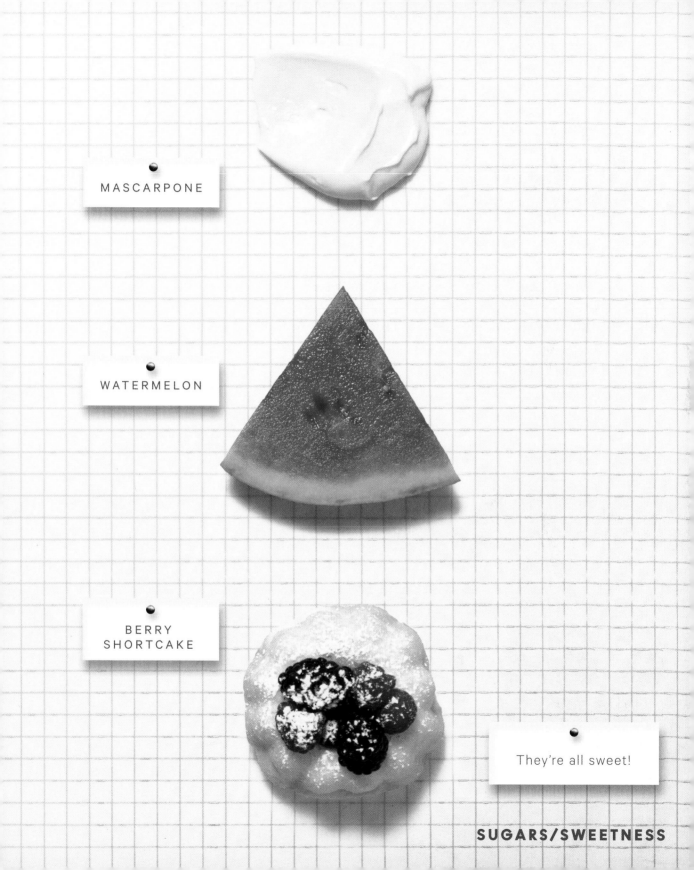

MASCARPONE

WATERMELON

BERRY
SHORTCAKE

They're all sweet!

SUGARS/SWEETNESS

BROWNING

Browned food is delicious. It's why we like barbecue, crème brûlée, coffee, grilled onions, and s'mores—all of which would be impossible without sugars. Browning happens when sugars get so hot that they start to vibrate with energy. When they vibrate hard enough, they explode. Each microscopic piece of sugar shrapnel then explodes again, and those shards start smacking into one another, forming new compounds. This rippling cascade of reactions transforms sweet, colorless, odorless sugar into some of the deepest, most complex mixtures of taste, color, and aroma in existence. Browning is a sugar supernova.

There are two types of browning that can happen when we cook food: *caramelization* and *Maillard browning*. Both involve sugars exploding to produce dark flavors, but they require slightly different conditions to get started. Caramelization comes from pure sugars, while Maillard browning requires sugars *and* proteins. Since nearly all food contains at least a small amount of protein, true caramelization seldom happens apart from when we heat pure sugars to make caramel. The minute you add something as simple as butter to turn that caramel into toffee, the dairy proteins in the butter kick off a cascade of slightly different explosions: Maillard browning.

During Maillard browning, proteins (and amino acids, the links in a protein chain—see the Proteins chapter) act like lighter fluid, provoking the sugars to ignite more easily and contribute their own explosive flavors to the mix. The extra boost from proteins allows Maillard browning to happen with less heat than is necessary for caramelization.

Contrary to traditional pastry chef wisdom, however, neither style of browning has a specific starting temperature. Like all processes in this book, browning reactions depend on a combination of time *and* temperature (see the Heat chapter). We associate browning with grilling, frying, and other high-temperature techniques, but those temperatures are required only to make browning happen *quickly*. Browning can also creep along at much lower temperatures. Low-temperature browning may be too slow to be useful for cooking dinner, but it's a potent way to develop deep flavor over time. Tomatoes, raisins, and figs dried in the sun brown over the course of a few days. Balsamic vinegar, miso, and fish sauce take months to brown at the cooler temperatures of a cellar. Some of the sugars in our own bodies are browning right now, but at such a slow pace that we'll never notice it happening.

Caramelization comes from pure sugars, while Maillard browning requires sugars *and* proteins.

Another important difference between the two types of browning is that caramelization can happen with any sugar, but Maillard browning won't work with sucrose. Sucrose has a hard time getting together with proteins because of its structure, so it needs to be broken apart into glucose and fructose before the protein fuse can be lit. This means that the table sugar in a marinade, dough, or sauce won't brown nearly as well as corn syrup, molasses, honey, fruit juice, milk, or any other source of non-sucrose sugars.

SUGARS / BROWNING

During caramelization, sugar molecules break apart
into a wide spectrum of different fragments, creating a
complex cocktail of taste, aroma, and color.

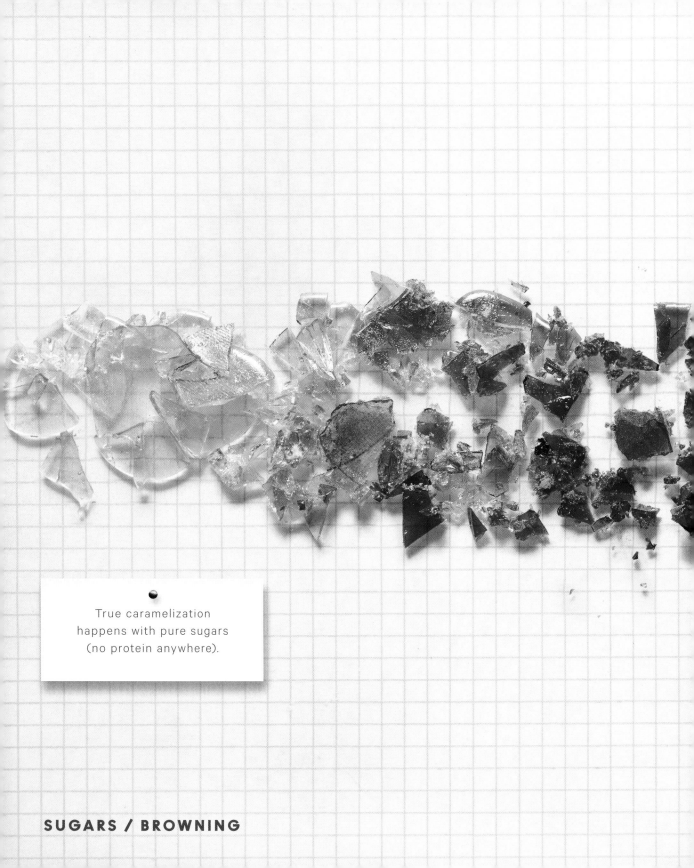

True caramelization
happens with pure sugars
(no protein anywhere).

SUGARS / BROWNING

CRYSTALLIZATION

We have a complicated relationship with sugar crystallization. In fudge and rock candy, crystals are the goal, but in toffee and ice cream, crystals are the enemy. Luckily, the rules of crystallization are simple.

Like freezing water, sugar crystallization is all about organization. Crystallization transforms a swarming throng of dissolved sugars into evenly spaced and neatly organized rows to form a perfect crystal. Each type of sugar crystallizes pure, and everything else gets pushed to the periphery. If there are too many impurities, they get in the way of the sugar molecules as they try to form ranks, and the crystal won't form.

Using impurities to prevent sugar crystallization is the rationale behind most successful candy recipes. Chewy caramels, toffee, and hard candies like lollipops and Jolly Ranchers require boiling sugar and water down to thick concentrated syrup. With so little water to keep the sugar dissolved, these mixtures are perfect breeding grounds for crystals, which would ruin the texture and glossy appearance of the candy. Adding multiple sugar types creates a chaotic mess of mismatched puzzle pieces, thwarting the order required to form crystals. This is why most candy recipes also call for a mix of sugars such as table sugar (sucrose) and corn syrup (glucose) or honey (fructose and glucose). Other Ingredients like carbs, proteins, and lipids can get in sugar's way as well, which is why we add cream, butter, fruit purees, and starches to keep sauces, candy, jams, and pastry fillings smooth and crystal-free.

Putting Ingredient roadblocks in sugar's way isn't the only way to prevent crystals from forming. We can also control how baby crystals are born. All crystals

> Like freezing water, sugar crystallization is all about organization.

start from a seed, which can be anything from the side of a pan or an air bubble to the wire on a whisk or an undissolved speck of something. Like an oyster starting with a grain of sand to create a pearl, the seed gives the sugars a place to congregate. Using a clean cooking vessel, ensuring that everything is evenly dissolved, and avoiding gratuitous stirring are effective ways to keep crystals from starting. The size, shape, and material of the cooking vessel also matter, since hot and cold spots can jump-start the crystallization process. Hot spots can scorch sugary liquids to create a dry crust full of seeds, and cold spots provide a calm refuge for crystals to grow. Even the best chefs struggle with making caramel when the pan is the wrong size for the burner.

Once crystals have seeded, the way they grow depends on temperature and agitation. When the mixture is hot, it's hard for crystals to form because the sugars are excited and bouncing around. As the mixture cools, the sugars calm down enough to organize themselves into crystals. Slow cooling gives crystals plenty of time to grow, and quick cooling keeps growth to a minimum. Agitation keeps things small by breaking up crystals as they form, turning large crystals into seeds for smaller ones. Combining temperature and agitation allows us to control the size of the crystals, from big and coarse to small and fine. When we leave rock candy undisturbed or stir and stretch fudge, fondant, and taffy as they cool, we are farming crystals of exactly the size that we want to create a desired texture. We want large, beautiful crystals to dissolve slowly in the mouth in the case of rock candy, and small crystals so tiny that they are almost imperceptible in smooth, creamy fudge; malleable fondant; and chewy taffy.

Combining temperature and agitation allows us to control the size of the crystals, from big and coarse to small and fine.

SINGLE TYPE OF SUGAR

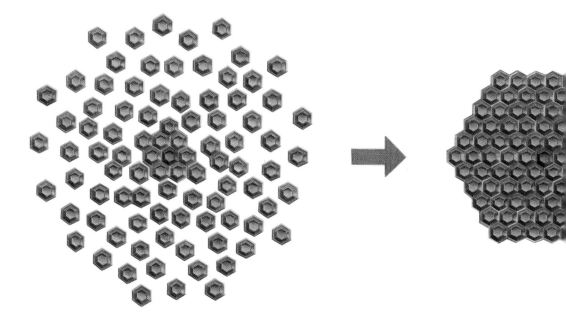

OTHER SUGARS ADDED

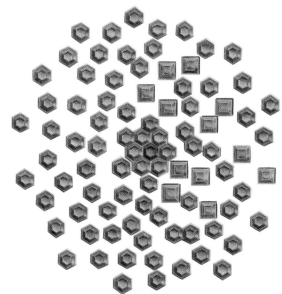

SUGARS / CRYSTALLIZATION

Concentrated sugars of a single type will form crystals around a central "seed" unless we add alternative sugars to break up the organized structure and keep the mixture smooth.

ROCK CANDY

CARAMEL SAUCE

ROCK CANDY
A stick left undisturbed in concentrated sugar syrup will form big crystals.

ROCKY ROAD FUDGE
———
BLACK-AND-WHITE COOKIE
Fudge and icing stirred during cooling form lots of small crystals.

PARMESAN
Crunchy crystals form as water evaporates during aging.

CARAMEL

Lipids and proteins from cream get in the way of sugars to prevent crystallization.

HONEY

If crystallized, it can be heated to melt crystals.

CANDIED APPLE

SUCKERS

Corn syrup (glucose) is added to keep table sugar (sucrose) from crystallizing, allowing a brittle glass to form.

SUGARS / CRYSTALLIZATION

DISSOLVING

When sugar dissolves in water, several molecules of water mob each molecule of sugar. A mutual attraction ensues, and both sugar and water are held together, almost like being trapped in a binding force field. In that force field, water is spoken for—it puts all of its energy into the relationship and is less likely to get into any shenanigans with other molecules.

Free, unfettered water aids and abets activities that we often want to discourage in food. It allows microbes to grow and flourish, which can make food spoil and even become dangerous if left unchecked. Water also helps proteins move around and form webby networks that can become coarse and tough. Sugar helps put an end to the mischief. Dissolved sugar jealously hoards water, keeping it away from microbes, which is why preserves are preserved. Sugar also robs proteins of the water they require to glob together, keeping meringues from becoming grainy, custards from curdling, and cakes from becoming tough.

In addition to making water less interested in other Ingredients, sugar latches on to water to help water maintain its physical form. Sugar makes it harder for water to crystallize. Bound to sugar, water freezes at lower temperatures with smaller, slower-growing ice crystals. This helps us fine-tune the texture of ice cream, sorbet, and any other frozen food. Too little sugar gives a coarse, icy texture, and too much sugar makes a thick slush that won't freeze properly. Sugar also makes it harder for water to evaporate. Adding sugar to a batter, dough, brine, or any other mixture prevents water from escaping during cooking and while on the shelf. Using sugar to bind water is one of the secrets to preserving moisture in food, whether in a molasses cookie, birthday cake, dried date, or chicken leg. Minerals are the other Ingredient best suited to preserving moisture in food (see the Minerals chapter). We don't enjoy salt concentrations of more than a couple of percent in our food, however, so we usually rely on sugar in most situations that require a lot of water binding.

Free, unfettered water aids and abets activities that we often want to discourage in food.

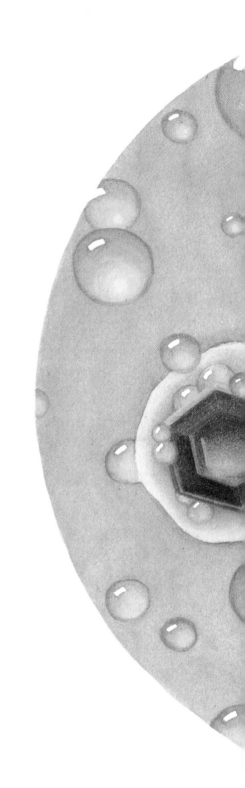

SUGARS / DISSOLVING

Sugars bind water, preventing it from evaporating, freezing, interacting with other Ingredients, and helping microbes grow.

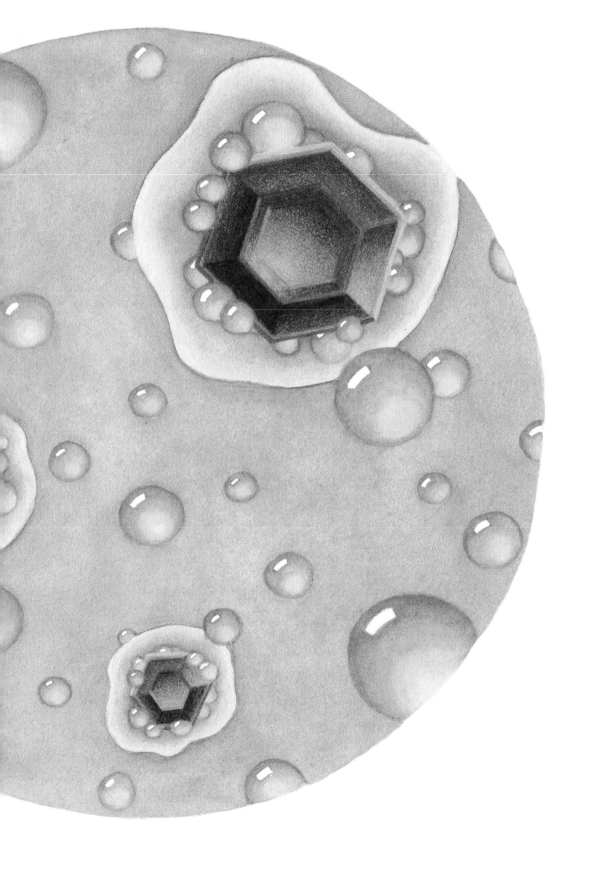

CHICKEN LEG
Sugar in brine holds on to water for moist meat.

MOLASSES COOKIE
Sugar in dough holds on to water for moist cookie.

MANGO POP
Sugar makes water form smaller crystals to prevent grittiness.

SUGARS / DISSOLVING

MERINGUE

CUPCAKE

Sugar binds water to prevent egg and wheat proteins from forming tough, gritty networks.

CANDIED BACON

RASPBERRY JAM

Lots of sugar steals water from microbes, preserving food.

THICKENING

Sugar isn't particularly big, so a single sugar molecule doesn't present much of a roadblock to water, especially when compared to carbs and proteins. We often eat food that is more than half sugar by weight, however, and all of those sugar molecules add up. In large quantities, sugar slows everything down.

Thickening water with sugar is good for more than just creating sticky syrups. When included as part of a greater recipe, the sugar-thickened water acts like cement, sealing cracks and reinforcing delicate structures. Meringues, marshmallows, and the foam on a beer contain delicate scaffolding made of tiny gas bubbles trapped in water. Pure water is thin, so it quickly drains off that scaffolding and weeps in the bottom of the cup or bowl. Like honey slowly oozing out of a jar, sugar-thickened water takes much longer to weep out, prolonging the life of the foam. This works for gels as well. While sugar lacks the long, noodle-y structure of proteins and carbs that is required to make gel networks by itself, it can help seal cracks in the gel. This sugar cement keeps water from seeping out of everything from custards and jams to cheese and Jell-O.

With enough sugar, water can get so thick and syrupy that it stops moving completely. Everything gets so jammed up that neither water nor sugar crystals can form because nothing can

> When included as part of a greater recipe, the sugar-thickened water acts like cement.

move enough to organize into crystal grids. These glassy traffic jams shatter just like real glass, adding texture and intrigue to crème brûlée, hard candy, glazed hams, Peking duck, and the coatings on everything from expensive bonbons to M&M's.

SUGARS / THICKENING

Sugars get in the way of water to thicken liquids, but they're small, so it takes large amounts of sugar to make a difference. High enough concentrations will create a glass if the sugars are prevented from crystallizing.

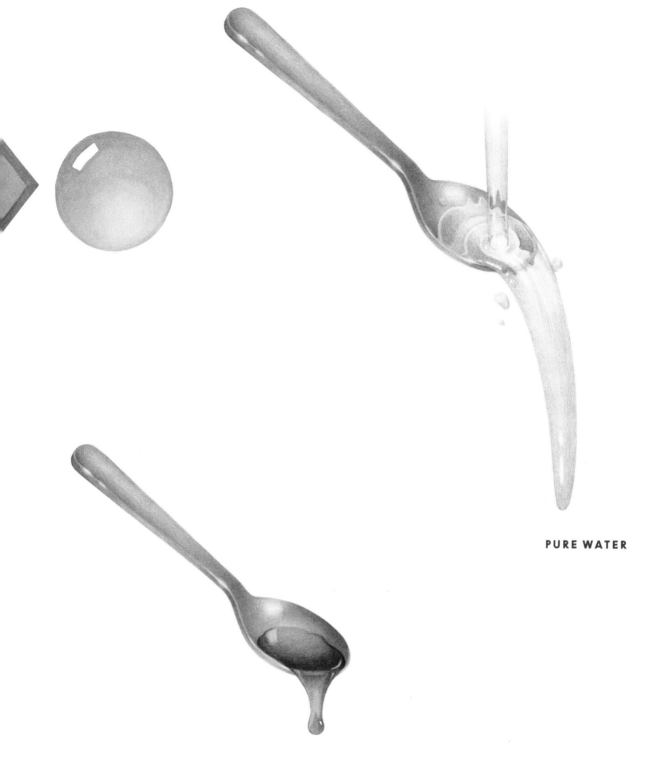

PURE WATER

HONEY

AGAVE NECTAR

MOLASSES

BLUEBERRY COMPOTE

SUGARS / THICKENING

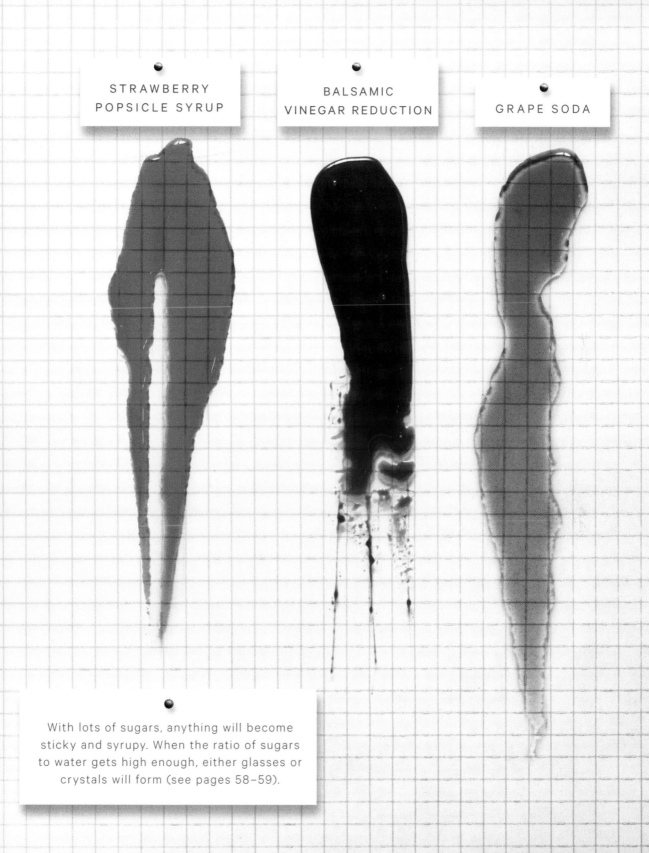

STRAWBERRY
POPSICLE SYRUP

BALSAMIC
VINEGAR REDUCTION

GRAPE SODA

With lots of sugars, anything will become
sticky and syrupy. When the ratio of sugars
to water gets high enough, either glasses or
crystals will form (see pages 58–59).

FERMENTATION

Sugar works the same way for microbes as it does for humans: We thrive on sugars but only in moderation. We need some sugar to survive, but a sixty-four-ounce Big Gulp is good for nobody's health. Too much sugar stunts microbe growth because it deprives microbes of water, but a small amount is necessary for them to survive. In fact, it's usually their main source of food.

Microbes eat sugar . . . messily. Microbes don't have mouths, so they eat by wrapping themselves around sugar molecules and smashing the sugar through their flabby membranes into their gooey interior. Once inside, their internal machinery chops the sugars into pieces. Microbes use some of the sugar pieces as fuel to move, grow, and reproduce; anything they can't use gets thrown back onto the food. Microbes are sloppy eaters, but luckily for us, the leftover sugar scraps they discard can be delicious.

In terms of flavor, sugar can be boring—it's purely sweet, with neither color nor aroma. But scraps of sugars left over from a microbial feeding frenzy are never boring. Depending on the type of microbe and its appetite based on temperature, competition, and environment, these pieces of sugar can take a lot of delicious forms. A lot of microbes turn sugar into acid, adding tang to fermented dairy, charcuterie, pickles, and vinegar. Others are

Too much sugar stunts microbe growth because it deprives microbes of water, but a small amount is ncecessary for them to survive.

moonshiners, surreptitiously creating alcohol under the cover of darkness. Some microbes turn sugar into gas, which leavens dough and makes sparkling wine, beer, and even some pickles feel fizzy on the tongue. Occasionally, microbes use sugar as the main entrée in a multicourse feast that includes other Ingredients like proteins, carbs, and lipids. As scraps of other Ingredients join the mix, the fermentation process becomes more pungent and aromatic. These scraps accumulate over time, which is what turns simple things like grape juice and milk into wines and cheeses with flavors so complex that we still haven't completely figured them out.

Occasionally, microbes use sugar as the main entrée in a multicourse feast that includes other Ingredients like proteins, carbs, and lipids.

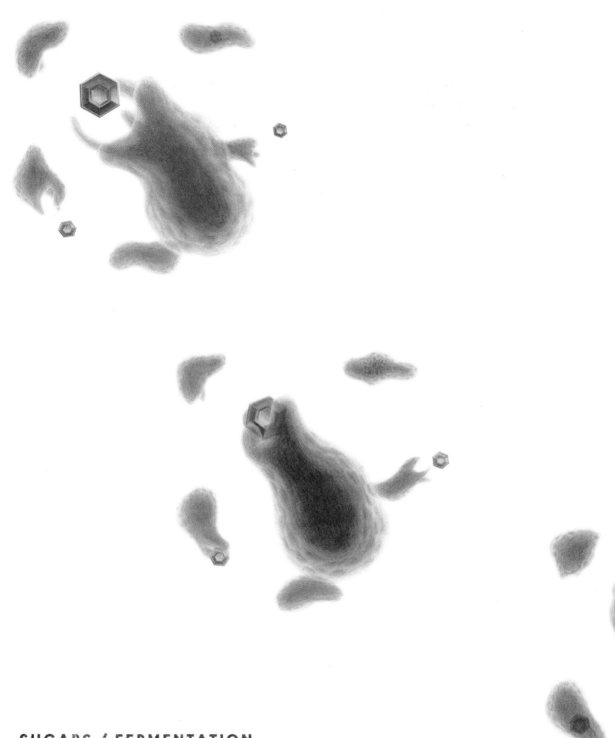

SUGARS / FERMENTATION

In most types of fermentation, microbes absorb sugars, break
them apart for energy and building materials, and discard unused
fragments, which can be delicious.

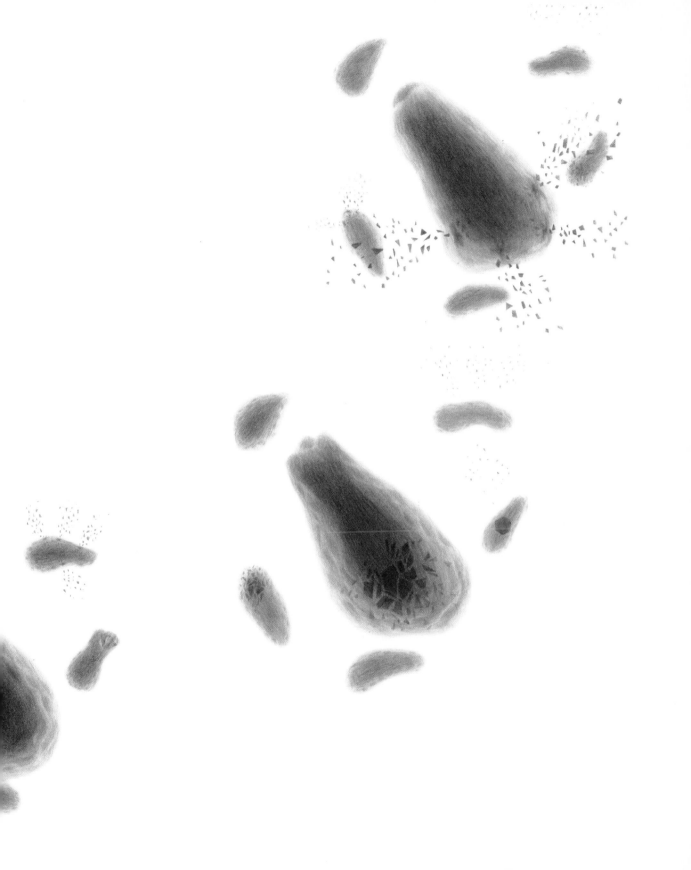

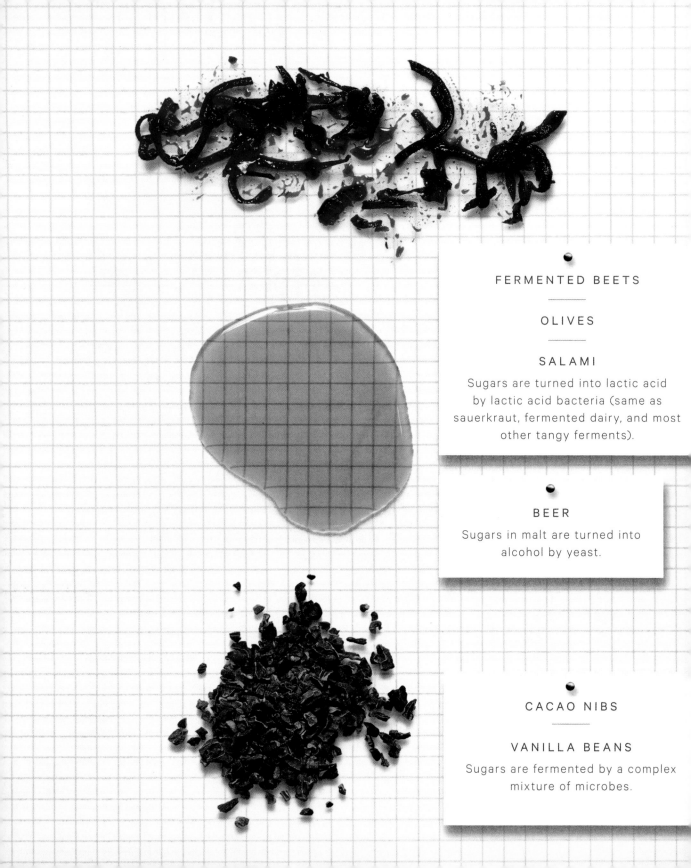

FERMENTED BEETS

OLIVES

SALAMI

Sugars are turned into lactic acid by lactic acid bacteria (same as sauerkraut, fermented dairy, and most other tangy ferments).

BEER

Sugars in malt are turned into alcohol by yeast.

CACAO NIBS

VANILLA BEANS

Sugars are fermented by a complex mixture of microbes.

HATCHO MISO

Molds kick-start the fermentation
process by liberating free sugars.

SUGARS / FERMENTATION

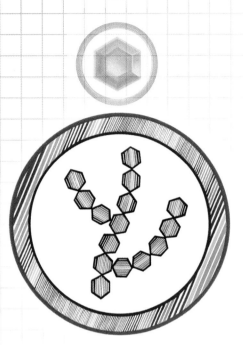

ink

CARBO

HYDRATES

Carbohydrates are made of sugars (which are technically a tiny type of carb themselves)—anywhere from a few dozen to a few thousand sugars strung together to form long chains. They come in a variety of clumsy, tangled shapes that ensnare the other Ingredients, giving us crispy potato chips and thick tomato sauce. Their convoluted structure helps them play five basic roles in food:

- They **DISSOLVE**.
- They **THICKEN LIQUIDS**.
- They **FORM GELS**.
- They **BIND UP TASTE AND AROMA COMPOUNDS**.
- They **BREAK DOWN INTO FREE SUGARS**.

DISSOLVING

The bigger an Ingredient is, the more difficult it is to dissolve. Water has an easy time prying apart small Ingredients like minerals and sugars, but it struggles when untangling knots of gigantic carb chains. In fact, water can make things worse by swarming over the chains on the outside of the clumps, getting stuck, and sealing the rest of the carbs inside.

To get carbs to do anything useful, like thickening a sauce or gelling a jelly, we need to help water break up the lumps *before* it gets in its own way. This is why recipes call for whisking starch into cold water before adding it to broth to make gravy. Using cold water slows everything down and buys us extra time to separate the individual carbs before lumps form. The other approach is to pre-blend carbs with other Ingredients like lipids and sugars to keep the chains separated from one another and evenly dispersed. This is the idea behind blending flour with butter or oil in roux before making gumbo and sifting pectin with sugar before making jam.

Once each chain has been separated and exposed, most carbs need heat to dissolve completely. Carb chains are so long and convoluted that water needs an energy boost from heat to wiggle into all the nooks and crannies. This is why flour, pectin, and agar need to be boiled vigorously after whisking to dissolve. There are some exceptions to that rule—carbs such as xanthan gum and modified starches have been pretreated to allow easy access to water, so they hydrate as soon as the chains are exposed. Each type of carb has its own particular set of conditions that it needs to dissolve, and these basics give us the keys to using any of them.

Like sugar and minerals, dissolved carbs hold water's attention, preventing it from associating with microbes, forming frozen ice crystals, or escaping into the air as steam. Water is less fascinated by carbs than it is by sugars and minerals, however, so the effect isn't

Water has an easy time prying apart small Ingredients like minerals and sugars, but it struggles when untangling knots of gigantic carb chains.

nearly as strong. Swarms of captivated water molecules surround sugar from all angles, so each sugar molecule binds a lot of water. The sugary links in a carb chain, however, are packed close together with no room for water molecules in between, so the chains bind water poorly. The starch in a baguette traps some water, but the loaf will dry out much faster than a sugary cake. Starchy sauces freeze harder and with coarser ice crystals than sugary ice cream. Pasta, crackers, and other carb-rich foods without a lot of sugar need to be dried to about 10 percent water to keep microbes down, while sugary jams can be shelf stable even at 50 percent water.

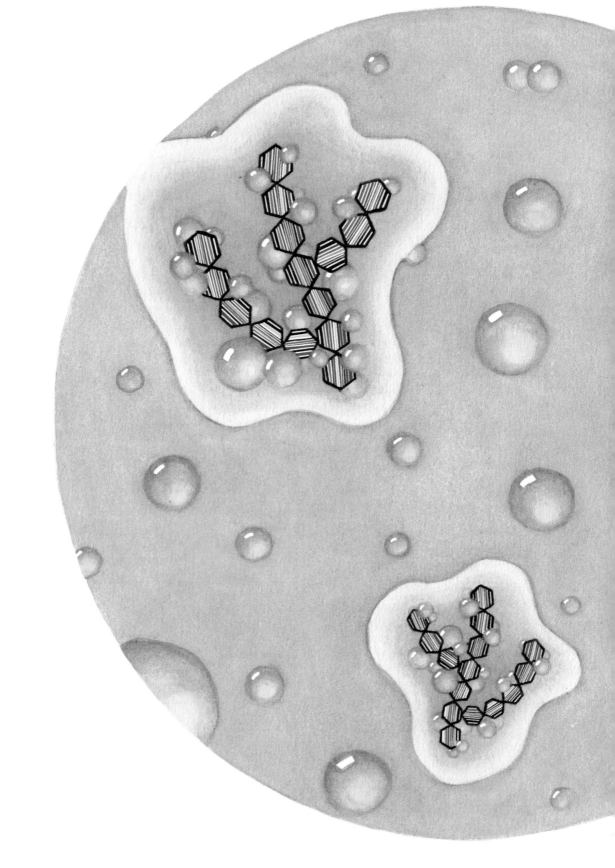

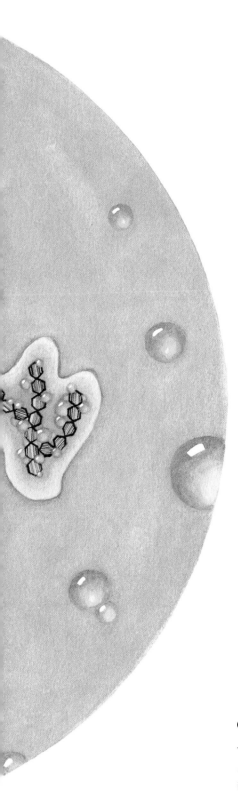

CARBOHYDRATES / DISSOLVING

When carbs are properly dissolved, each molecule is surrounded by its own entourage of water molecules. Because water can't fit in between the links in carb chains, carbs are less effective at binding water than sugars.

OKRA

STEWED OKRA

ORANGE

ORANGE
MARMALADE

CARBOHYDRATES / DISSOLVING

CARROT

CARROT PUREE

For carbs to be effective thickeners, emulsion stabilizers, or anything else, they have to be properly dissolved. When working with whole ingredients, this is accomplished by cooking, stirring, and/or blending.

APPLE

APPLESAUCE

THICKENING

Swimming in an open pool is simple. Unimpeded, it's easy to move straight ahead. Swimming through a field of kelp and seaweed is complicated. The long, tangled vines turn that straight course into a winding ordeal. This is how carbs make liquids thicker: They get in the way of water as it flows.

A single pile of seaweed is easy to swim around, and a clumpy ball of carbs won't do much to thicken a liquid. As we discussed in the Water chapter, the better distributed an Ingredient, the more obstructions it causes. Whisking, blending, or otherwise evenly distributing carb chains helps maximize their thickening power.

Plants are the best place to look for carbs. They have neither bones nor muscles, so fruits, vegetables, legumes, grains, spices, and herbs rely on carbs like starch, pectin, cellulose, and a bunch of others for structure, movement, and energy.

Thickening food with carbs from plants started as a crude, brute-force technique, but we've come a long way from squashing fruit into pastes and smashing roots with a rock for caveman mashed potatoes. The trick to thickening with plants is figuring out how to open the natural package in which the carbs come. Trapped inside a clove of garlic, a chickpea, a fig, or a squash, carbs can't do much to thicken a liquid, but by simmering or pureeing to draw them out, we unleash their thickening potential. In some cases, we use tons of heat and the physical abuse of a blender to make perfectly smooth, homogeneous beet soup or eggplant puree. Other times, we want to thicken a liquid without completely destroying the

Carbs are the most powerful Ingredients for thickening, which also makes them the best Ingredients for crispiness.

thickener. Good risotto, tapioca pudding, and steel-cut oatmeal are about balance—we want to lure some carbs out into the liquid to make it creamy while leaving enough carb scaffolding in place to preserve the structure and chewy texture of the solid pieces.

Carbs from whole ingredients provide a lot of thickening power, but they come adorned with taste, aroma, color, and other Ingredients that may not be ideal for every application. In situations where we want precise thickening *and nothing else,* refined carbs are the way to go. Cornstarch, potato starch, arrowroot, and other powdered carbs like agar, xanthan gum, and pectin are pure carb chains, extracted from grains, roots, fruit, seaweed, and even some microbes. Each type of purified carb has different lengths and shapes of chains, which gives them a range of thickening power and requirements for dissolving.

As we discussed in the Water chapter, crispiness happens at the extreme end of thickening. Carbs are the most powerful Ingredients for thickening, which also makes them the best Ingredients for crispiness. Potatoes, onions, green bananas, and anything else with a lot of carb chains become glassy and crisp when fried, baked, toasted, or dried to remove water. Natural carbs aren't the only route to crispy success: Food that lacks enough carbs of its own can be breaded or dredged with any carb-rich substance from starch and flour to breakfast cereal, ground spices, dried vegetables, and crushed tortilla chips to create a shattering crust.

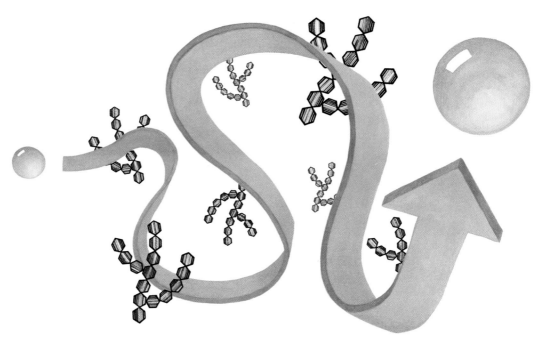

CARBOHYDRATES / THICKENING

Carbs get in the way of water to thicken liquids. Their long
structure makes them the most effective Ingredient for
thickening. With enough carbs, water will stop moving completely
to form a glass, but a gel will not form unless carb chains overlap.

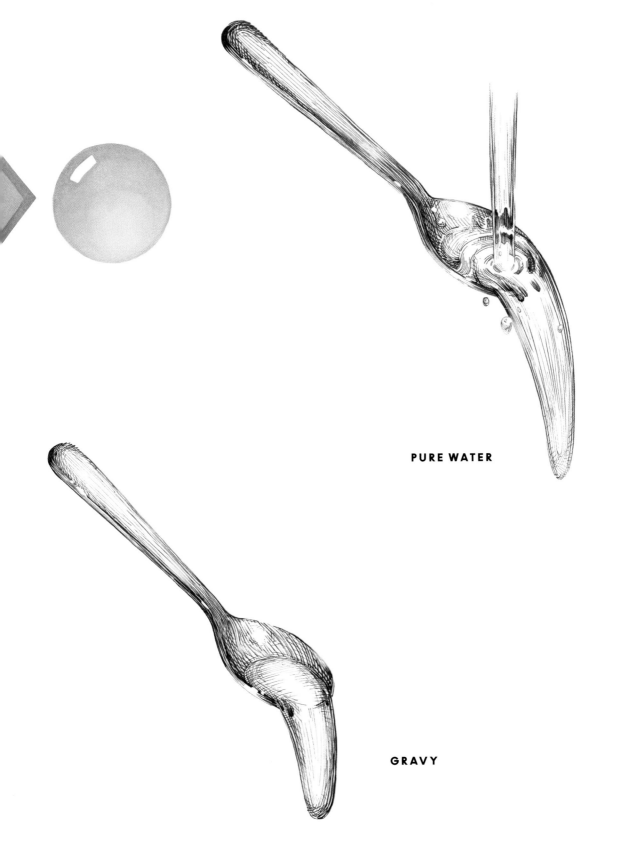

PURE WATER

GRAVY

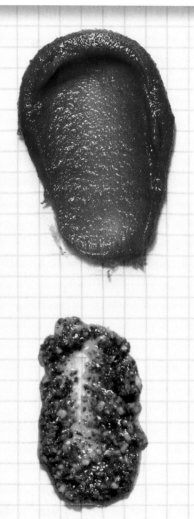

TOMATO PASTE
Pectin from tomato cell walls

PESTO
Starch from crushed nuts and
cellulose from herb leaves

BAKLAVA
Crispy glass made of wheat starch

CARBOHYDRATES / THICKENING

RANCH DRESSING

Xanthan gum, extracted and purified from bacteria

SEAWEED CRACKER

Crispy glass made of algae carbs

ORANGE JUICE

Pectin from orange rind and pulp

RICE CAKE

Crispy glass made of rice starch

GELLING

When used as thickeners, carbs work like roadblocks, delaying water as it tries to flow. When used to create gels, individual carb chains fuse together to lock water in a cage and stop it completely.

As with dissolving and thickening, evenly dispersing carbs to avoid clumping is the first step to gelling. Once the carbs are nicely unpacked, getting a gel to form means assembling the carbs into a cage. No matter how crowded the carbs get, a liquid won't gel unless the chains crisscross and link up to create a 3-D network. Each type of carb needs a special set of circumstances to stitch that network together. Pectin for jam, gumdrops, and marmalade need the help of low pH and a bunch of sugar. Some carbs like alginate, carrageenan, and gellan require minerals, which work like rivets to precisely weld individual chains together into a gel network, creating the dazzling array of spheres, cubes, and other innovative shapes dreamed up by contemporary chefs. Most starches will gel if you simply heat them up and let them cool back down, which is how pecan pie, pancakes, gnocchi, and glass noodles work.

The journey from thick to gelled isn't always a one-way trip. Some foods hang out in a gray area between thick liquid and gel, and they can be temporarily thinned out by jiggling, squeezing, or stirring. The

No matter how crowded the carbs get, a liquid won't gel unless the chains crisscross and link up to create a 3-D network.

agitation breaks the fragile gel cages open, letting everything flow until the cages re-form. The most extreme examples of this gray area are bottled salad dressings and ketchup. The tangled carbs stubbornly hold everything in the bottle until the tipping point, where the cage breaks open. Suddenly, the condiments gush out of the bottle, magically transforming from a solid to a liquid. As soon as they hit the plate, your clothes, or the floor, these "fluid gels" firm up again as the carb cage snaps back into place. We see some of the same patterns in foods like yogurt and panna cotta, but those are thickened with proteins (see the Proteins chapter), which share some personality traits with carbs, as they are also shaped like long chains.

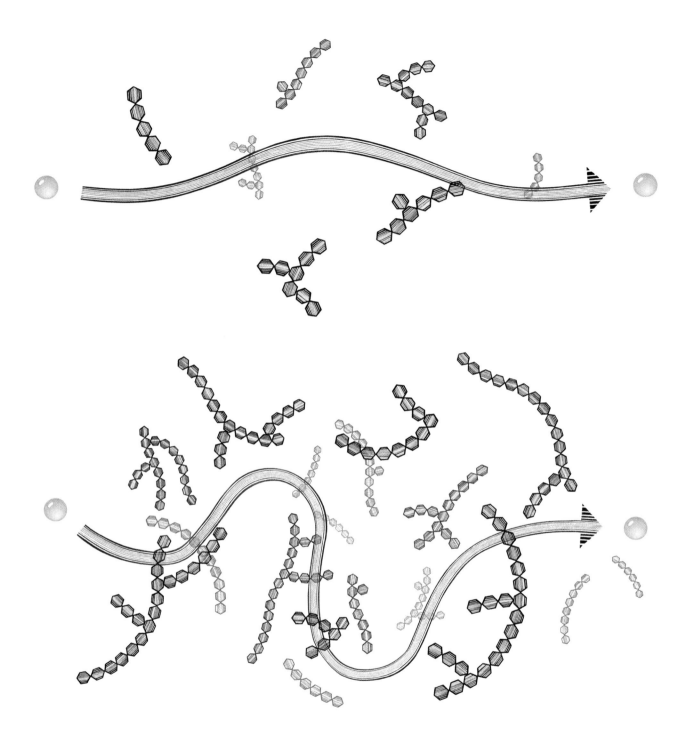

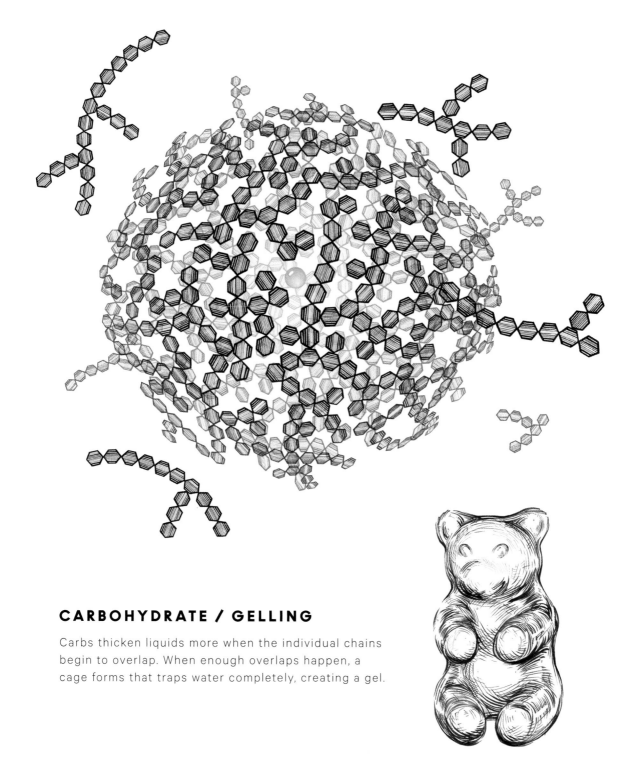

CARBOHYDRATE / GELLING

Carbs thicken liquids more when the individual chains begin to overlap. When enough overlaps happen, a cage forms that traps water completely, creating a gel.

GUMMY BEAR

GLASS NOODLE
Mung bean starch gels noodles without help from proteins.

SPANISH LIME
———
TOMATO
Flesh is held together by pectin.

PANCAKE
Wheat starch works with gluten proteins to create coordinated gel.

CARBOHYDRATES / GELLING

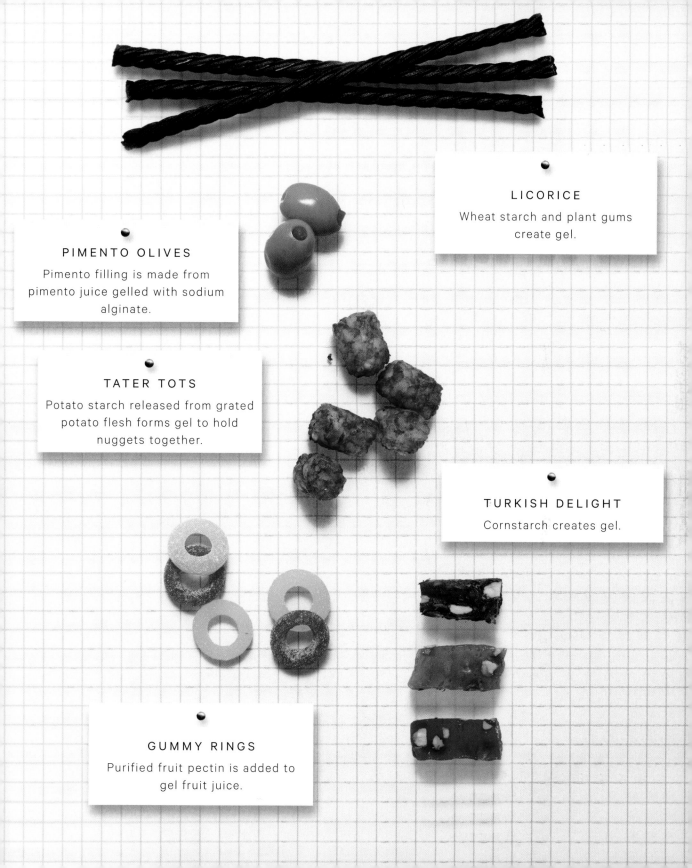

LICORICE
Wheat starch and plant gums create gel.

PIMENTO OLIVES
Pimento filling is made from pimento juice gelled with sodium alginate.

TATER TOTS
Potato starch released from grated potato flesh forms gel to hold nuggets together.

TURKISH DELIGHT
Cornstarch creates gel.

GUMMY RINGS
Purified fruit pectin is added to gel fruit juice.

TASTE + AROMA BINDING

A tangled forest of carbs will do more than change the texture of your food—it will also affect how food tastes and smells. Aroma and taste compounds get wrapped up in the carb chains, unable to move. In your mouth, they can't escape onto your tongue or float up into your nose, so the flavor feels muted. Carbs can be black holes for taste and aroma.

Of all the carbs, starch is the most aggressive binder of flavor, which can cause starchy food to seem lackluster. If the flavor of chicken stock, shrimp, or onion puree isn't strong enough, then the addition of starch will make the resulting sauce, fried shrimp, or risotto almost tasteless. This is one of the reasons that restaurant chefs in recent years have turned to other carbs such as pectin, agar, xanthan gum, and gellan gum to manipulate the texture of food. These alternative thickeners and gelling agents bind fewer flavorful compounds and leave the color of food unchanged, allowing delicate flavors to shine through. A strawberry sauce thickened with starch can be pretty weak, but that same sauce thickened with xanthan gum can shine with berry goodness.

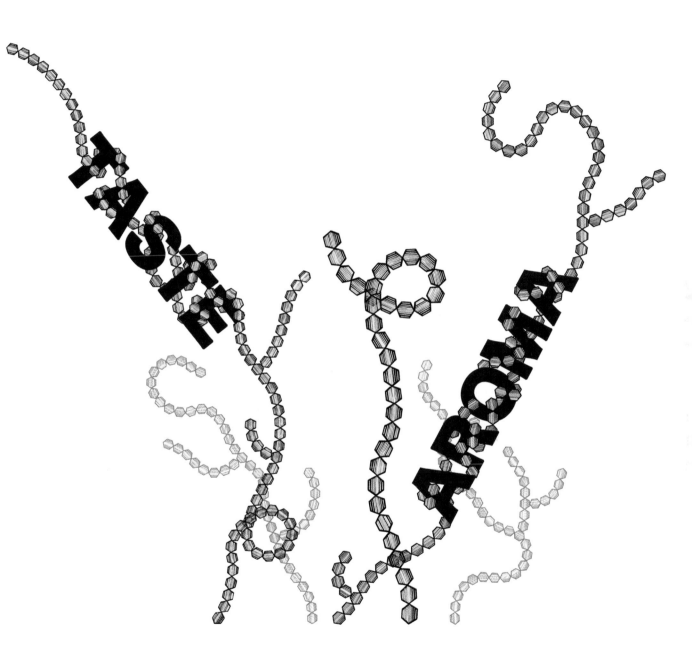

CARBOHYDRATES / TASTE + AROMA BINDING

Taste and aroma can get entangled in carb chains which
prevents them from escaping onto your tongue or into your
nose, muting the flavor of food.

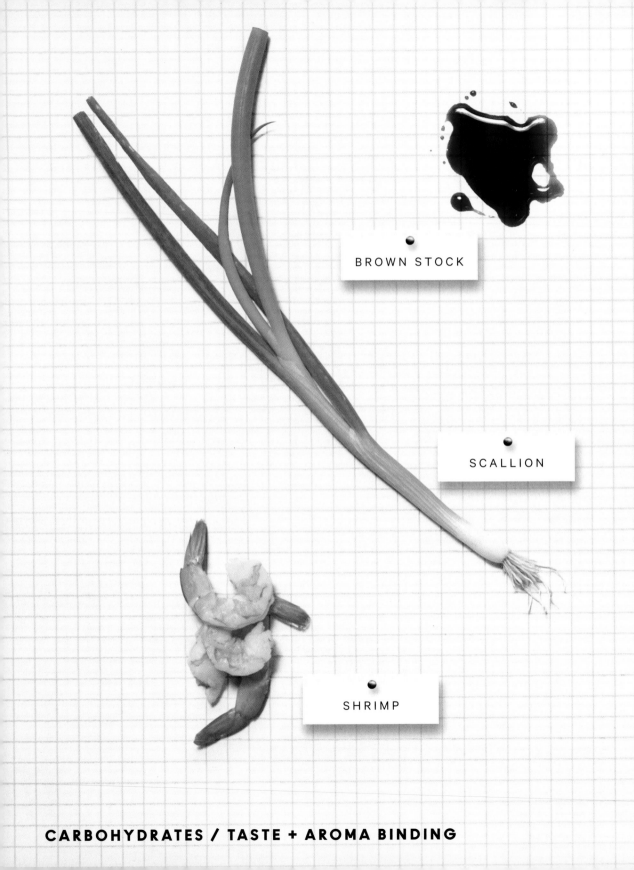

BROWN STOCK

SCALLION

SHRIMP

CARBOHYDRATES / TASTE + AROMA BINDING

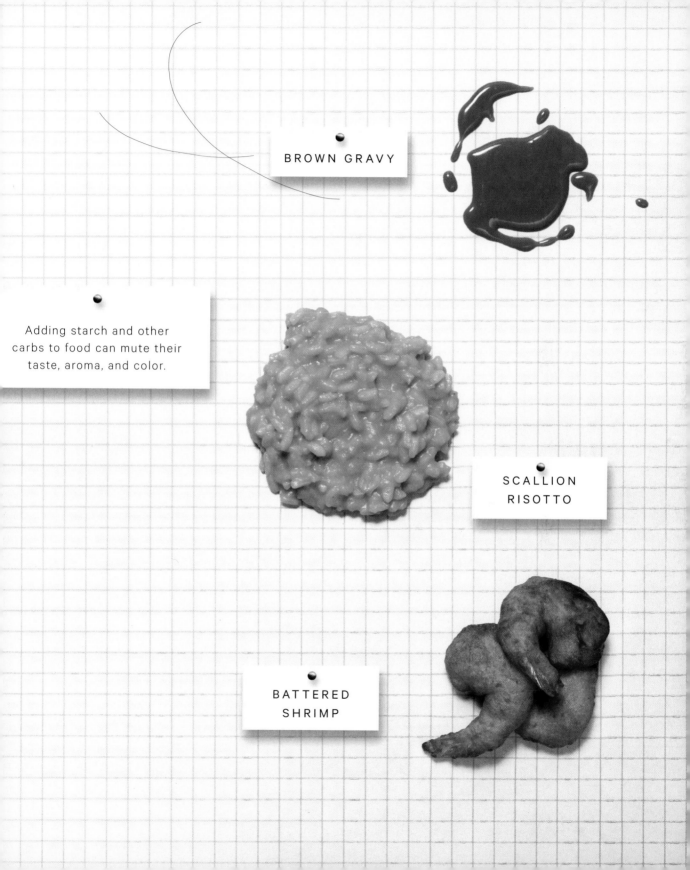

BROWN GRAVY

Adding starch and other carbs to food can mute their taste, aroma, and color.

SCALLION RISOTTO

BATTERED SHRIMP

BREAKDOWN

Carbohydrates are chains of sugars. When the chains break, they start to act more like individual sugars.

With time and heat, carb gels can come apart at the junctions between chains, and the chains themselves can break into smaller pieces. This is how most plant foods become softer during cooking. The carb scaffolding "backbone" in each cell of a carrot or a leaf of kale starts to come apart, and the tissue transitions from crisp and hard to pliable and soft. Cooking pries carb chains apart enough to tenderize vegetables, but breaking them into individual sugars usually requires the help of enzymes.

Enzymes, mini protein machines (see the Proteins chapter), act like tiny molecular knives for chopping carbs into sugary bits. Enzymes that digest carbs are found everywhere from the cells of plants to microbes and the guts of animals. This means that sweet potatoes, koji (rice mold used to make soy sauce and miso), and honeybee intestines share something in common: They all have enzymes with the ability to turn carb chains into mounds of free sugars.

Unlike the carb chains that it came from, that liberated sugar is sweet. For something to be tasted, it needs to be able to fit into your taste buds. A chain of a million sugars sewn together is a big, heavy mess. Taste buds can't grab on to it, so it has no taste of its own. As fruits ripen, sweet potatoes age, miso ferments, and barley malts, they become sweeter. Sweet peas are one of the weird exceptions to that rule; peas have enzymes that work backward, knitting sugars together to form carb chains.

As soon as peas are harvested, these enzymes kick in, transforming them into starchy, tasteless pellets.

Carb breakdown affects more than sweetness. As chains break, they brown better, bind more water, ferment more quickly, crystallize more easily, and lose their ability to thicken and gel. Finding the balance between the two sides of the spectrum is key. Long-fermented bread dough browns more deeply in the oven, and ripe plantains are more caramel-y when fried, but both become softer and more difficult to keep intact from the lack of carb structure. Old potatoes turn golden brown more quickly in the fryer, but the lack of thickening power makes them less crispy. Ripe figs are great for sweetening desserts, but green figs are potent thickeners that can make fillings and frostings more stable. Sweet, ripe grapes need to be turned into wine immediately, but starchy grains can be brewed whenever the brewer decides to break those starches down into malt. The journey from long carb chains to individual sugars lies on a continuum, and finding the right balance between the two is key for creating food with the right texture, taste, and color.

For something to be tasted, it needs to be able to fit into your taste buds.

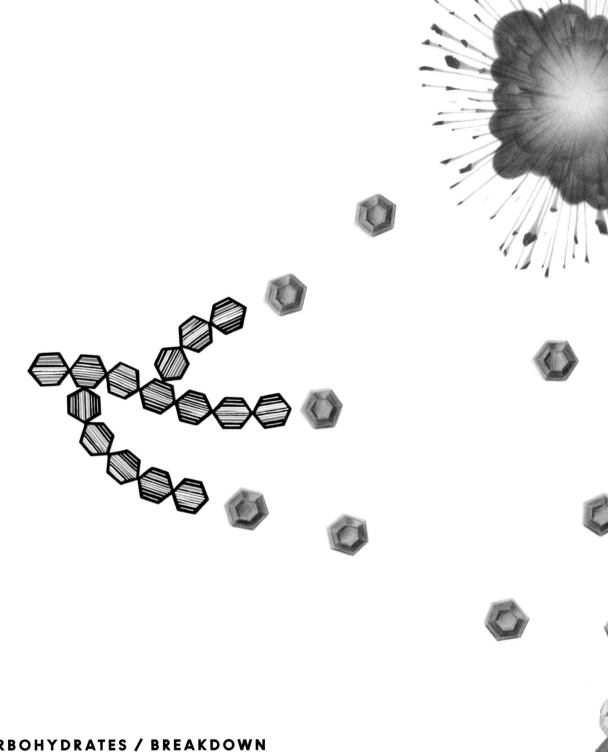

CARBOHYDRATES / BREAKDOWN

Carb chains can break apart to yield free sugars, which exhibit all of the traits of sugars, including browning, water binding, fermentation, sweetness, and crystallization.

SUGAR CRYSTAL

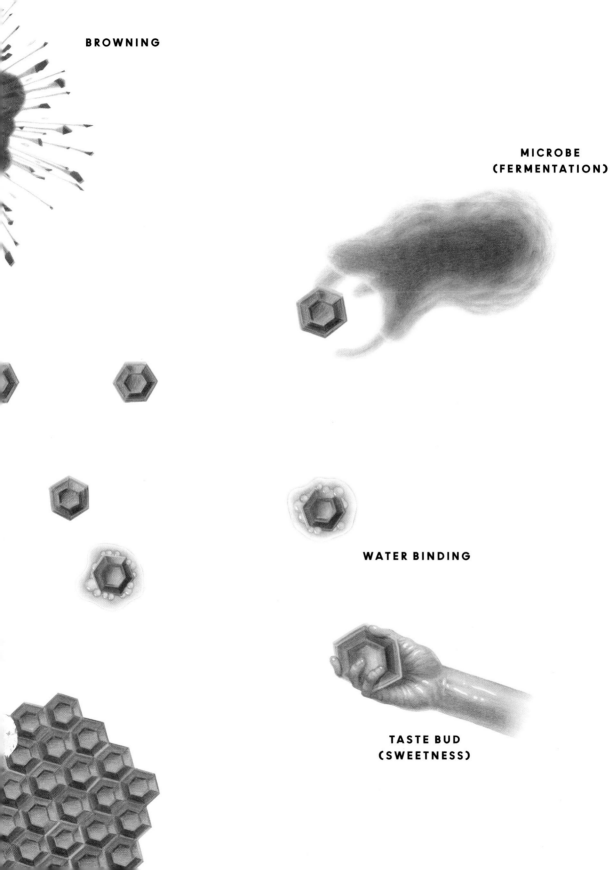

BROWNING

MICROBE
(FERMENTATION)

WATER BINDING

TASTE BUD
(SWEETNESS)

SWEET POTATOES
Natural enzymes make them sweeter during low-temperature cooking.

MALTED BARLEY
Sugars from starch breakdown allow fermentation and use of malt in sweet desserts.

POMEGRANATE
Starch, pectin, and other carbs are broken down during ripening of all fruit.

CARBOHYDRATES / BREAKDOWN

PERSIMMONS
Starch, pectin, and other carbs are broken down during ripening of all fruit.

SWEET MISO
Fermenting at temperatures that favor breakdown of carbs results in more sweetness than savoriness.

PLANTAINS
When green, they're as starchy as potatoes, but when fully ripe, a third of their weight is free sugar.

oil paint

LIPI

DS

Lipids don't play nice with water. They are greasy, oily, and fatty. They also smell. Lipids are sensitive to harsh conditions like exposure to heat, light, and air, which can skew their smelliness from floral to fishy in a flash. Here are the defining traits of lipids:

- They **FORM EMULSIONS**.
- They **ACT AS EMULSIFIERS**.
- They **STORE COMPOUNDS THAT HATE WATER**.
- They **CRYSTALLIZE AND MELT**.
- They heat to **HIGH TEMPERATURES** without evaporating.
- They **BREAK DOWN** into flavorful pieces.

EMULSIONS

Most lipids and water hate each other. These lipids do everything in their power to get away from water. If they are forced into contact with water and have no possibility of escape, the best they can do is band together. Like penguins in Antarctica hiding from the cold, the lipids crowd together in a giant mob so that only those on the outside have to be uncomfortable. Water and lipids coexist in nearly everything we eat, so they have to find some way to share the same space. Emulsions—tiny droplets of oil in water (or vice versa)—are the answer.

The main goal of making an emulsion is to keep the oil droplets well mixed and separated from one another. This isn't easy. Lipids fight against being split up because banding together helps them remain as far from water as possible. The first step toward longer-lasting emulsions is to make the droplets really, really small. This extends the life of the emulsion because it takes a while for all of the pieces to stick back together. There are no immortal emulsions—all emulsions break eventually—but all we care about is keeping emulsions intact long enough to eat them. A vinaigrette for a family dinner can be whisked together seconds before serving, but a vinaigrette for restaurant service needs to stay together for hours. To allow for these time windows, we use larger amounts of blending power to blitz the droplets into smaller and smaller bits. A high-powered blender makes a better emulsion than a hand blender, which is better than a whisk, which is better than a fork, which is better than a spoon, which is better than a finger.

Small droplets are a good start for an emulsion that's built to last. After that first step, preserving an emulsion is all about managing the interaction between droplets. Our tool belt for keeping emulsified droplets apart includes stabilizers, emulsifiers, and temperature control. Stabilizers are anything that makes water thicker, including carbs, proteins, and any other Ingredient that gets

Water can become so crowded with lipid droplets that they can't help but bump into one another.

in the way of water. From the lipids' point of view, stabilizers make it harder to swim to the nearest lipid droplet ally. Lipids are also less dense than water, and stabilizers make it harder for droplets to float up and congregate at the surface of the emulsion. Emulsifiers are anything that keeps droplets from being able to unite. Most food emulsifiers are either lipids or proteins. They coat the surface of droplets so that if two droplets successfully rendezvous, they bounce off each other rather than joining together. The last trick for preserving an emulsion is turning down the temperature (see the Heat chapter). Removing heat makes everything move slower, and emulsions last longer when lipid droplets can only slowly trudge toward one another.

Even though lipids don't typically dissolve in water, they can still act like roadblocks in its path. Emulsions thicken liquids because water has to navigate around each of the lipid droplets like greasy traffic cones. This is what happens when we swirl butter into a sauce to give it velvety thickness or make mayonnaise so stiff that it will hold up a spoon. Like the other Ingredients, lipids thicken liquids best when they are evenly distributed. It takes much less time for water to avoid a few big droplets than to weave through a minefield of thousands of them, so fine emulsions are not only more stable than coarse ones; they're thicker, too.

There is a limit to the thickening effect, however. Water can become so crowded with lipid droplets that they can't help but bump into one another. When this happens, the emulsion becomes more fragile and more likely to break. We then see little oil puddles start to form on the surface, heralding the impending doom of the emulsion. Adding more water gives the lipid droplets more room to breathe. A few drops of lemon juice, stock, milk, or anything else that contains water can help bring overcrowded beurre blanc, hollandaise, lemon curd, and mayo back from the brink of breaking.

LIPIDS / EMULSIONS

To form an emulsion, we force lipids and water into contact by creating droplets, but the droplets want to glob together to separate out. Starting with small droplets, removing heat, and adding emulsifiers and/or stabilizers can help prolong an emulsion's life.

Lipid droplets can get in water's way, thickening liquids as long as the emulsion is intact.

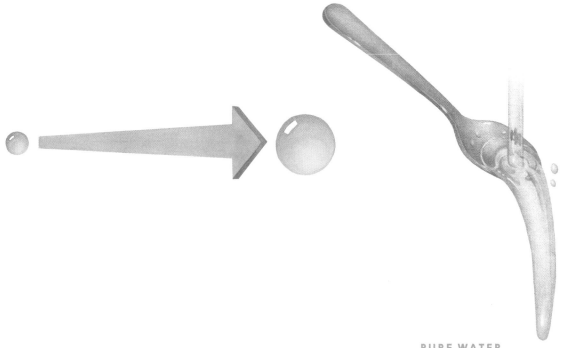

PURE WATER

BUTTER SAUCE

CHOCOLATE GANACHE

Cocoa butter emulsion preserved by carbs in cacao pulp and added lipid emulsifiers

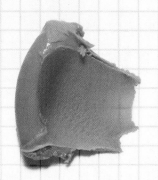

PEANUT BUTTER

Peanut oil emulsion preserved by carbs and proteins in ground peanuts and added emulsifiers

GUACAMOLE

Avocado oil emulsion stabilized by carbs in avocado pulp

DIJON MUSTARD
Mustard oil emulsion
preserved by proteins and
carbs in seed hulls

GLAZED RADISH
Emulsion of milk fat from butter
whisked into white wine

WASABI MAYONNAISE
Canola oil emulsion preserved with
protein and carbs from egg yolks
and wasabi powder

BRIOCHE
Emulsion of milk fat from butter
evenly distributed throughout
dough, preserved by proteins
and carbs

EMULSIFIERS

Oil droplets want to band together to flee from water. The critical keystone of their plan is the moment when two droplets merge to become one. Emulsifiers get in the way of that union. Emulsifiers are lipid chaperones, preventing droplets from getting inappropriately close.

Emulsifiers are molecules with one part that loves water and one part that hates it. Emulsifiers work because they are hybrids—the water-loving part hangs out in water, and the oil-loving part buries itself in oil. The end result is a droplet studded with bumpers on its surface. Those bumpers make droplets pinball off of one another rather than join together. Some specialized lipids, like lecithin and cholesterol, are emulsifiers that can be found in a lot of places, from garlic and egg yolks to dairy products and vegetable juices. Emulsifiers can also come from common lipids, like fats and oils, that get broken apart into fragments, exposing the combination of water-hating and water-loving parts that allow them to straddle the border between watery and oily worlds. Used fryer oil is full of emulsifiers from lipids that have broken into pieces over several minutes of prolonged heating. These emulsifiers help intact lipids in the oil get closer to the water-rich food they are cooking, which is why food browns and cooks better in "broken in" oil than

Emulsifiers are lipid chaperones, preventing droplets from getting inappropiately close.

in fresh oil. Apart from lipids, there are dozens of foods with proteins that also work as emulsifiers (see the Proteins chapter).

In general, it's a good idea to add an emulsifier to food *before* trying to make an emulsion. That way, each droplet gets assigned its own personal detail of emulsifier chaperones as soon as it is made. This is why mayonnaise recipes usually start with egg yolks, vinaigrette recipes start with mustard, and traditional Spanish aioli recipes start with crushed garlic. In an emergency, however, adding an emulsifier to an already-formed emulsion in need of some help can still be an effective fix, as long as you stir or blend it well enough to disperse the emulsifiers to all the parts of the emulsion where they are needed.

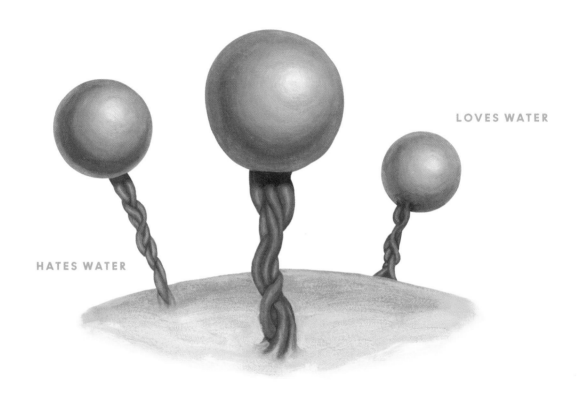

LOVES WATER

HATES WATER

LIPIDS / EMULSIFIERS

Emulsifiers are any molecule with one part that loves
water and one part that hates it.

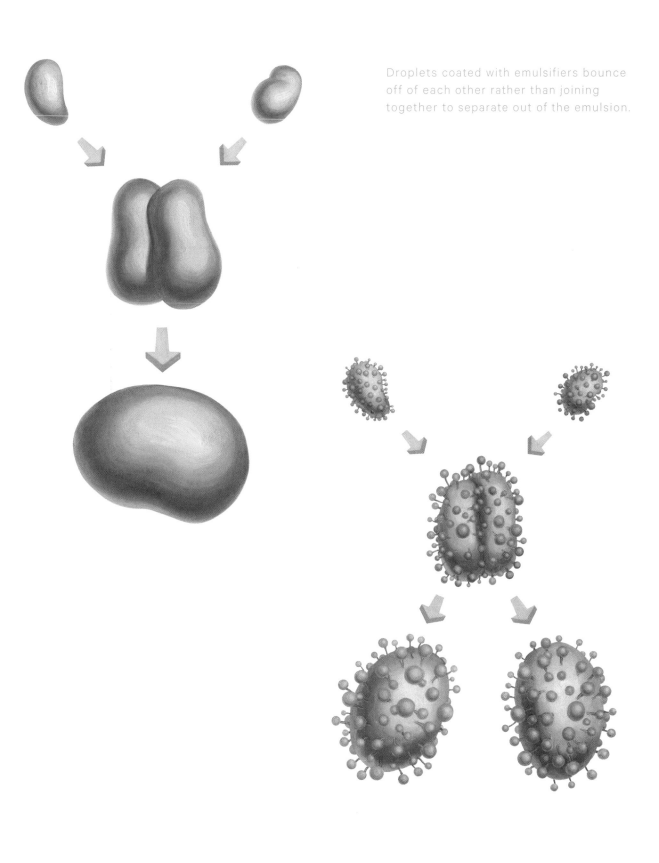

Droplets coated with emulsifiers bounce off of each other rather than joining together to separate out of the emulsion.

ROASTED
GARLIC

COCONUT

All of these contain lipids. Those
lipids form natural emulsions that
are preserved by lipid emulsifiers
present in the food.

EGG YOLK

AVOCADO

PINK
PEPPERCORNS

MUSTARD
POWDER

LIPIDS / EMULSIFERS

STORAGE OF WATER-HATERS

A lot of colors and aromas hate water. Most food is made of water, which puts those colors and aromas in an awkward situation and often ends in those compounds wasting away or escaping the food altogether. Lipids act like a safe haven for colors and aromas, keeping them in place long enough for food to reach the mouth.

There is an old chef adage: "Fat has flavor." Since taste compounds generally prefer water to lipids, an obnoxiously precise version of that adage would be "Fat has aroma." Most aroma compounds love to lounge around in lipids. This is why curry oil is a thing and curry water is not. Food without lipids is usually bland and lacking fragrance because aroma can't stick around long enough for us to enjoy it. Fat-free cream cheese is terrible. Culinary students who are taught the French technique of skimming all of the fat from their stocks and broths, yet to also add aromatic ingredients like parsley and thyme, are receiving very mixed messages. Japanese ramen chefs, on the other hand, swirl loads of fat into their broths, providing a comfortable home for the aromas of toasted garlic and spices.

We need to be careful about making aromas too comfortable, however. Food with too many lipids can wrap up the aroma too tightly. This can create a similar effect to carbs binding taste and aroma—food can pass through your mouth and be down your throat before aroma has a chance to escape into your nose. Adding lipids to food is a balancing act. Some lipids are helpful for imbuing food with fragrant smells, but add too much, and they start to suck aromas out of the air (and by extension, your nose).

LIPIDS / STORAGE OF WATER-HATERS

Lipids can serve as storage for things that hate water in our food, including many important aromas, colors, and nutrients.

CREAM CHEESE

Fat-free cream cheese may get close to the texture of regular, but the lack of lipids changes the way it holds on to creamy aromas.

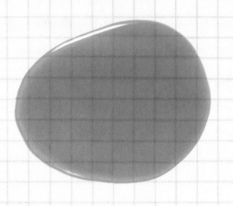

FENNEL OIL

The bright green color comes from chlorophyll, which dissolves in lipids along with the aroma of fennel.

PROSCIUTTO

Nutty, meaty, funky aromas hang out in the abundant fat of the ham.

LIPIDS / STORAGE OF WATER-HATERS

HERBED CHICKEN

Aromas from the herbs perfume the chicken— more are captured in the fatty skin and dark meat than the lean breast meat.

TOASTED SESAME SEEDS

The aroma of toasted nuts, seeds, and spices are held in their natural oil reservoirs.

RED PALM OIL

The red-orange color comes from carotenoids, the same lipid-friendly pigments in carrots and sweet potatoes.

LEMON ZEST

Most of the aroma of citrus fruit is stored in oil pockets in the outer skin.

MELT +
CRYSTALLIZE

All lipids melt and crystallize with the addition and removal of heat. Controlling those transitions helps us make food that is flaky or crumbly, dense or light, creamy or greasy.

Fats are the most common type of lipid in food, and they come in two varieties: saturated and unsaturated. One of the biggest differences between the two is their melting temperature, which largely depends on their shape. Fat molecules are shaped like sticks. Saturated fats are perfectly straight sticks, which allows them to be packed closely together like sardines. That close packing makes saturated fats like lard, duck fat, and cocoa butter dense and solid, so it takes more heat to melt and pry them apart than unsaturated fats. Unsaturated fats have bends and kinks that jut out like elbows in their shape. It's really hard to dance closely with somebody who keeps elbowing you in the face, and unsaturated fats pack together less closely than saturated fats for the same reason. This makes things like fish oil, olive oil, and canola oil melt more easily than saturated fats. In general, the word *oil* usually refers to fats that are liquid at room temperature.

Lipids have more complex shapes than water and

If lipids crystallize quickly, they melt differently than lipids that crystallized slowly.

sugar, so their molecules can snap together in different orientations to form crystals of several different shapes. If lipids crystallize quickly, they melt differently than lipids that crystallized slowly. Good chocolate is a study in lipid crystals because cocoa butter can form six different shapes of crystals, and we enjoy type number five more than the others. These crystals make chocolate that is glossy, snappy, and melts in your mouth, not in your hand. To make those crystals, we put chocolate through a tempering process. Tempering is a pattern of raising and lowering the temperature of the chocolate to make the lipids dance and shift in a way that leaves them settled down in our favorite crystal orientation. If chocolate melts in your pocket and resolidifies, the carefully curated crystals disappear, and the texture changes. The snappy texture goes away, and the chocolate becomes dull and more prone to melting all over your fingers. The same thing can happen with butter. Butter left out of the fridge too long becomes grainy and hard when put back in the cold.

When multiple crystal types exist at the same time, food feels greasy. The lipids in olive and canola oil tend to favor a single crystal type, so they melt quickly

> When multiple crystal types exist at the same time, food feels greasy.

at a single temperature point. Beef tallow and pork fat are made of a mixture of several crystal types, so they melt gradually over a range of temperatures. Rather than a quick transition from solid to liquid, these lipids go through a phase that is somewhere in between—a journey through the land of grease.

Since lipids and water hate each other, Ingredients that dissolve in water have a hard time traversing blobs of lipids in food. This makes lipids useful for controlling how stuff moves throughout food. Depending on the solidity of the lipids, they can either partially break up groups of water-loving things or keep them separated completely. In particular, there are a lot of culinary techniques geared toward controlling proteins, especially gluten, with lipids. Breaking up gluten's ability to form interconnected chewy networks is what makes croissants flaky, piecrust crumbly, and brioche tender. The difference among the textures of those three is a result of how the lipids are distributed. Solid fat forms firm borders between layers of dough in croissants and around pebble-sized balls of dough in piecrust. Brioche dough features liquid fat distributed evenly throughout the dough to create a general tenderizing effect. Any lipid could be used in all three

Since lipids and water hate each other, Ingredients that dissolve in water have a hard time traversing blobs of lipids in food.

scenarios as long as it is the right consistency. Olive oil croissants and duck fat brioche would work perfectly well as long as the croissants are made in a cold room with solid olive oil and the duck fat is melted prior to adding it to the brioche dough.

BUTTER

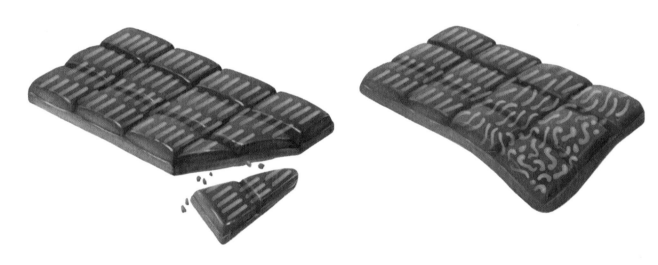

TEMPERED CHOCOLATE

UNTEMPERED CHOCOLATE

LIPIDS / MELT + CRYSTALLIZE

Crystallized lipids can create shiny, snappy textures in chocolate, or greasy textures in butter, depending on the orientation and type of crystals.

Saturated lipids can pack together more closely, solidifying more easily than unsaturated lipids, which are kept apart by their bent shapes.

LARD

CORN OIL

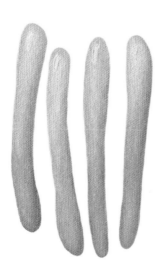

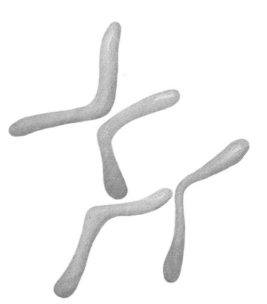

SATURATED LIPIDS

UNSATURATED LIPIDS

DANISH

Solid lipids are spread evenly between layers of dough to separate proteins into flaky sheets.

TEMPERED CHOCOLATE

Specific arrangement of lipid crystals creates shiny, snappy chocolate that melts at a temperature closer to that of your mouth than your hand.

COCONUT OIL

Solid at room temperature

LIPIDS / MELT + CRYSTALLIZE

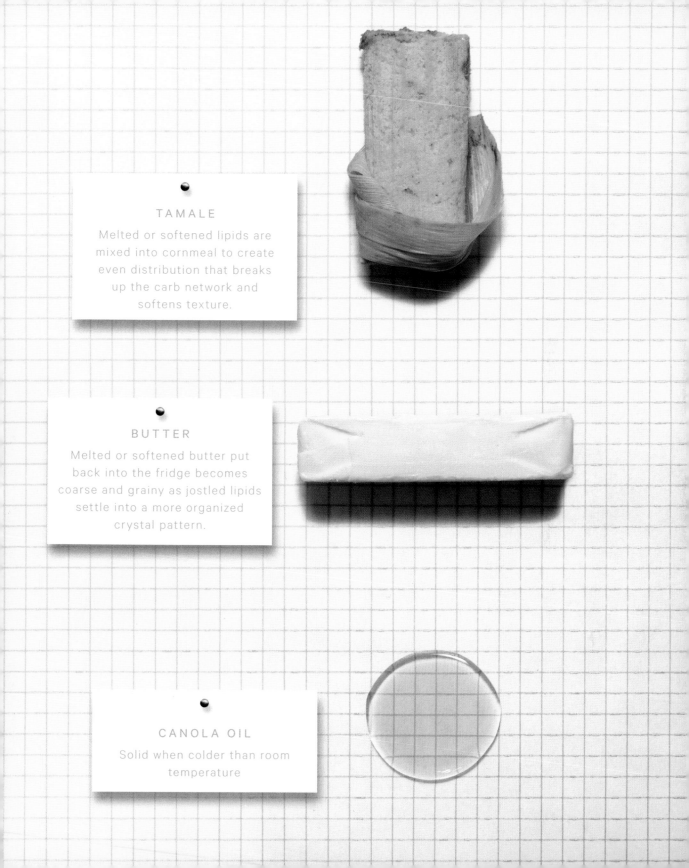

TAMALE
Melted or softened lipids are mixed into cornmeal to create even distribution that breaks up the carb network and softens texture.

BUTTER
Melted or softened butter put back into the fridge becomes coarse and grainy as jostled lipids settle into a more organized crystal pattern.

CANOLA OIL
Solid when colder than room temperature

HIGH HEAT

The best way to transfer heat evenly into food is to submerge it in hot liquid. In the kitchen, we have two very different types of liquid to choose from: lipids and water. Water does an admirable job in a lot of situations, but it has a serious limitation— without a pressure cooker, it can't get hotter than about 100˚C. Once it reaches that boiling point, water zips off into the air, taking heat with it and cooling your food. Lipids have crazy-high boiling points, which allow us to cook at much higher temperatures than in water.

Boiling requires being light enough to rocket off into the air, and fats and oils are heavy. They are so heavy that they will break down and burn before they boil. Submerging food completely in hot oil is one of the quickest ways to get heat in and water out of food. This is why nearly everything that comes out of a fryer is crispy and golden brown. Even when food isn't fully submerged, lipids act like heat-transfer middlemen, ushering heat toward food. This is especially useful when food has a weird shape or the heat source is inconsistent. Oil passes heat around like a bucket brigade, so as long as the surface of the food is covered, things tend to balance out. Brushing oddly shaped foods like mushrooms and peppers with oil for the grill or searing oblong fish

Lipids act like heat-transfer middlemen, ushering heat toward food.

fillets, chicken breasts, and beef ribs in a thin layer of fat in a pan makes things cook evenly.

All lipids that we use as cooking media have higher boiling points than water. Technically, we could use any of them for high-heat cooking, but saturated fats like beef tallow and coconut oil have an added advantage: They can handle high heat for a long time. Spending minutes to hours at temperatures in the hundreds of degrees is a taxing experience, and the more delicate, unsaturated fats tend to break down and burn, creating acrid aromas and off colors. Ignore the people on TV—you should *not* deep-fry those shrimp in forty dollars worth of extra virgin olive oil. In addition to some unsaturated lipids, EVOO contains various flavorful, heat-sensitive molecules leftover from pressing the olives, so if you must use olive oil for high heat cooking, go for a refined version instead.

Lipids are better than water for high-heat cooking because they can reach higher temperatures than water. Interestingly, lipids are also gentler on delicate foods than water when cooking at low temperatures because water holds more heat and conducts it more quickly than lipids at the same temperature. This means that if we had a pot of water and a pot of oil each at 70°C, food would cook quicker in the pot of water. Water has deeper pockets for storing heat than lipids, and the pipeline between heat and food is narrower through lipids than through water. For the most delicate foods, this allows us to do some really elegant

> The cooking medium doesn't vanish when the food is done cooking, so keep the serving situation in mind.

things. The technique of poaching lobster in butter at low heat (made famous at the French Laundry) is not only luxurious but also brilliant. Compared to pure water, butter transfers a small, slow trickle of heat to the lobster, cooking it more gently than would be possible in water at the same temperature. Cooking duck confit involves the same thought process. The goal is to cook duck legs low and slow in their own fat until they nearly fall apart with tenderness. Few things are lower and slower than a low-temperature pot of duck fat.

The cooking medium doesn't vanish when the food is done cooking, so keep the serving situation in mind. French fries can be cooked in beef tallow instead of peanut oil for more complex flavor, but those fries better be served good and hot. A soggy French fry cooked in canola oil isn't ideal, but at least it's tolerable. A soggy French fry coated in saturated, solidified, beefy grease is not a good time.

LIPIDS / HIGH HEAT

Lipids can be heated to much higher temperatures than water because of their higher boiling points, which keep molecules from escaping into the air and stealing heat from the food.

GRILLED CHEESE
Lipids in mayonnaise conduct heat to toast surface, boiling away water to encourage browning and crisping.

BUTTER-POACHED SALMON
Lipids in butter allow more gentle heat transfer than water at low temperatures for gentle cooking.

GRILLED ZUCCHINI
Oil brushed on surface helps compensate for uneven heat exposure during grilling.

LIPIDS / HIGH HEAT

SEARED TILAPIA
Lipids in cooking oil allow for high-heat cooking to brown outside before inside overcooks.

DUCK CONFIT
Lipids in duck fat allow more gentle heat transfer.

BRIOCHE BUN
Lipids facilitate even browning.

POTATO CHIP
Lipids help chips get hot enough to remove water quickly for maximum crispness.

BREAKDOWN

The majority of lipids in fats and oils are flavorless. They're too heavy to float into your nose, so they have barely any aroma; and they don't dissolve in water, so they have little, if any, taste. When broken into fragments, however, lipids are the smelliest Ingredients we have.

While some of the wide spectrum of food aromas comes from breakdown of sugars, proteins, and carbs, lipids are by far the smelliest because of how they break apart. They fall prey to enzymes, heat, oxygen, light, and even the presence of minerals. Those things jiggle, cut, burn, and break lipids apart into pieces. Each piece carries its own aroma, and together they form a mosaic of flavor with dozens of individual components. Lipid breakdown, like the caramelization of odorless sugar, showcases incredible complexity born of relatively flavorless beginnings.

With lipid breakdown, less is usually more. At low concentrations, pieces of lipids can smell great—they create hazelnut, lavender, cucumber, dried apricot, butter, pineapple, and other fruity, nutty, and grassy aromas. At high concentrations, they can be downright nasty. Many fishy, rotten, and murky odors come from runaway lipid breakdown.

Saturated fats made of short, stubby lipid molecules can take a beating. Red meat, dairy fat, and lard all contain substantial amounts of short, saturated fat molecules. Those sturdy molecules break apart slowly, shedding fragrant pieces in

When broken into fragments, lipids are the smelliest Ingredients we have.

a trickle rather than a flood. Aged red meat, cured hams, and aged cheeses are a testament to the pleasant, nutty funk that can come with controlled lipid breakdown.

Some lipids are more sensitive than others. The same bends in unsaturated lipid molecules that make them melt easier are also their weakest points. They are a lipid's Achilles heel—vulnerable to attack. Light, heat, and other agitators use these weak points to snap unsaturated lipids into pieces. The longer and more unsaturated a lipid molecule is, the more vulnerable it is, and the more rampant these changes become. Marine lipids, like those found in fish and shellfish, are some of the longest, most vulnerable lipids we can find out there. This is one of the reasons that seafood can go from pleasantly briny and fresh to overwhelmingly fishy so quickly. Marine lipids break down so easily, in fact, that we often associate their smell with any runaway lipid oxidation, regardless of the source of the lipids: Eventually, everything smells like fish.

All foods have lipids, not just the ones we think of as fatty. The aromas of melon, cucumber, and tomato come from lipid breakdown, and many of those molecules are the same as the ones that make fresh salmon smell green and grassy. Citrus fruits, herbs, spices, coffee, and other fragrant plant foods owe their intense aromas to essential oils. These cocktails of powerful-smelling lipid fragments dissolved in microscopic reservoirs of oil are stored in the fruits, seeds, leaves, and roots of these plants. Grating, grinding, and bruising them unlocks those reservoirs, providing us with a dazzling assortment of aromas to play with.

**INTACT
LIPID**

LIPIDS / BREAKDOWN

As lipids break apart, they can yield a whirlwind of aromas,
from pleasantly floral and fruity to unbearably fishy and stale
like cardboard.

PECANS

Unsaturated nut oils go rancid easily, producing musty, cardboard-like aromas.

MUSHROOMS

The characteristic scent of mushrooms comes from lipid-derived aromas.

LIPIDS / BREAKDOWN

HERB FLOWERS

Flowers communicate with us and each other using broken-down lipids as well as more complex molecules built from fragments of various Ingredients.

GOAT CHEESE

Lipid-digesting enzymes in goat milk give goat cheeses their funk.

SALMON

CUCUMBERS

Fresh salmon smells like cucumbers and melons before runaway lipid breakdown makes salmon smell fishy.

AGED BEEF SHORT RIB

Saturated lipids break down slowly, yielding pleasantly cheesy, nutty aromas.

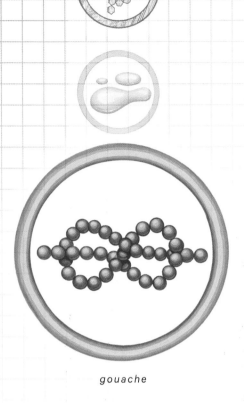

gouache

PRO

TEINS

Proteins are chains built from small, tasty links called amino acids, similar to how carbs are built from sugars. Carb chains flop around aimlessly, but protein chains fold themselves into shapes that make them come alive with activity. Proteins play important roles in everything from the obvious—eggs, bread, meat, and cheese—to the more surprising—apples, beer, onions, and soy sauce. We can understand what proteins do in any food by considering these basic rules:

- They **UNFOLD** and **COAGULATE**.
- They **DISSOLVE**.
- They work as **EMULSIFIERS**.
- They react with sugars to **BROWN** food.
- They act as **ENZYMES**.
- They **BREAK DOWN** into flavorful pieces.

UNFOLDING + COAGULATING

Proteins are made of long chains. The links in those chains are amino acids, small molecules that come in a variety of shapes and sizes. The chains of amino acids fold up like origami into a fully functional protein ready to do all kinds of stuff, from providing structural support to moving a muscle. There are about twenty different types of amino acids that function like letters in an alphabet. The order of letters dictates what word is spelled, and the order of amino acids determines what shape a protein takes.

Most of what proteins do results from their shape, and usually they take their shape for a simple reason: Some amino acids love water, and some hate it. If left to their own devices, the water-hating amino acids would blob together like lipids, putting as few of themselves in contact with water as possible. When they are locked into a protein chain, however, they have to come up with a different plan. The water-hating amino acids force the entire protein chain to curl up into a blob. They direct traffic so that all of the water-lovers stay on the outside of the blob and all of the water-haters hide on the inside.

Most proteins start out as nice, neat packages. Some cooking techniques seek to preserve those delicate bundles exactly as they are, while others use heat, pH changes, salt, and/or physical agitation to ruin that structure. Any time we whisk, chop, knead, boil, bake, fry, marinate, cure, or dry food, we change the shape of proteins.

When a protein unfolds, it exposes some of its water-hating interior, and those exposed spots panic. They run to

Any time we whisk, chop, knead, boil, bake, fry, marinate, cure, or dry food, we change the shape of proteins.

any nearby hiding place, often burying themselves in other exposed parts of neighboring proteins that are also panicking. Coagulation is when proteins stick together. How you unfold them affects how they coagulate.

Coagulation isn't an on-off switch. When we make proteins uncomfortable, they go through an arc of changes, from tightly folded to unfurled and floppy to interconnected gels to dense, gritty curds. Naturally folded proteins can thicken liquids but only slightly. A raw egg yolk won't do much to a sauce, and raw soy milk isn't very thick. As proteins begin to unfold, they take up more space. Unfurled proteins are bigger roadblocks in water's path, so they give us the velvety texture of silken tofu, cooked egg sauces, crème fraîche, and thick stews. If we continue agitating, heating, or curing, the proteins start to overlap and fuse. They create water-trapping cages to form a gel. This is where we get custards, ceviche, bread dough, meringues, and Jell-O. When pushed too far, the protein webs collapse on themselves. Water gets squeezed out. Grainy curds form. This can be great in the case of scrambled eggs and cheese curds, or it can be a sign of failure with weeping meringues, overcooked meat, and gritty quiche.

Proteins can also help things get crispy. Like carbs, proteins are very effective at locking up the other Ingredients, forming glassy structures when water is removed. Dehydrating protein-rich animal skin gives us pork rinds and crispy chicken, duck, and fish. Gluten lends starch a hand in creating crispiness in everything from the crust on baguettes to the breading on fried foods. It's even possible to make crunchy chips out of pure cheese, which has a lot of dairy protein.

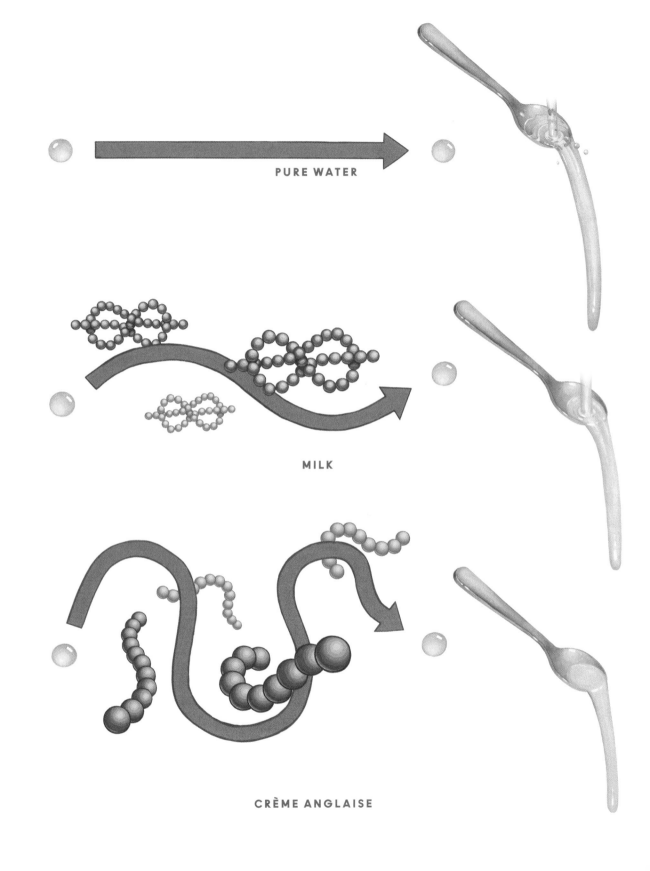

PURE WATER

MILK

CRÈME ANGLAISE

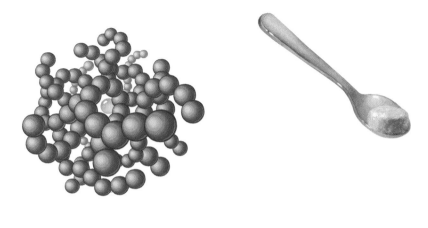

CUSTARD

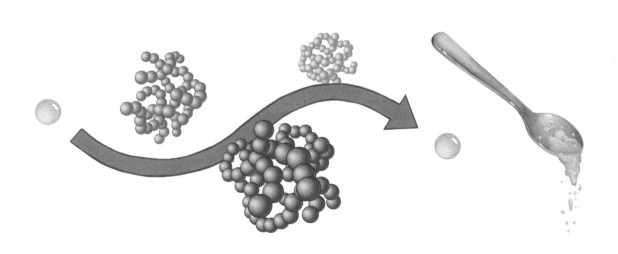

SCRAMBLED EGGS

PROTEINS / UNFOLDING + COAGULATING

Proteins get in the way of water, thickening liquids. They do so more effectively when slightly unfolded, and can even trap water completely to form gels. If overstressed, the gel network can collapse, forming compact, gritty balls of protein.

BURGER

Grinding and cooking causes proteins to unfold and set into a gel.

PARMESAN CRISP

When the ratio of proteins to water is high enough, crispy glasses form.

SCALLOP CEVICHE

Acid can be used to unfold and coagulate proteins.

DUCK PÂTÉ

BREAD CRUMBS

MOZZARELLA

PROTEINS / UNFOLDING + COAGULATING

DISSOLVING

Like carbohydrates, proteins can dissolve and hold water's attention, but not as well as sugars or minerals. Their structure means that water can only line up along the sides of a chain rather than completely surround each link. This limits the number of water molecules that can be held captive by each protein.

In spite of doing a relatively weak job, protein water binding is important for preventing water freezing, evaporation, and microbial growth in some situations. Egg yolks and egg whites added to ice cream and sorbet help prevent water from forming gritty ice crystals. Braised meat can be held hot for hours without drying out in part because dissolved gelatin makes it harder for water to fly away. Dairy proteins in cheese, meat proteins in aged charcuterie, and even soy proteins in fermented tofu also retain some moisture, even in products that have been aged for years. Concentrated proteins in energy bars and dried meat jerky help stave off microbes, but since they bind water only weakly, energy bars are usually full of sugar, and jerky is heavily salted to offer reinforcements in the battle to keep water away from microbes.

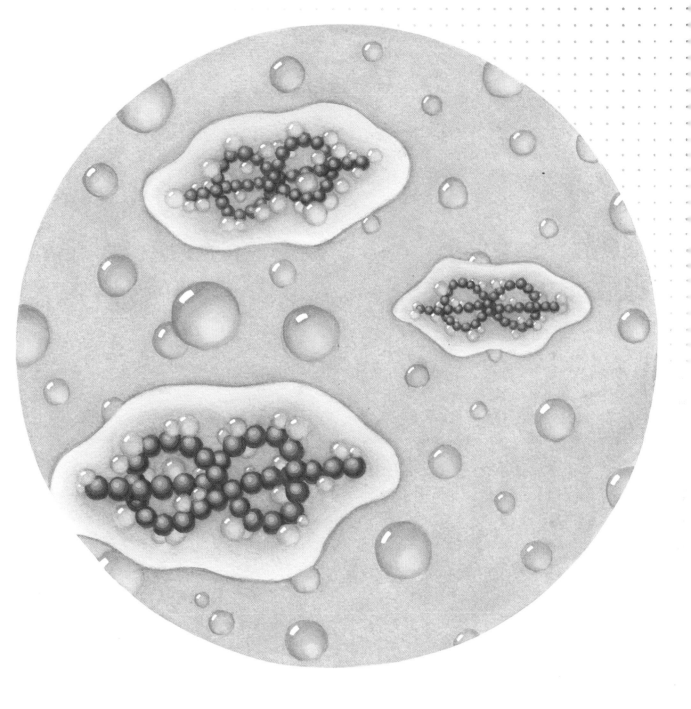

PROTEINS / DISSOLVING

Proteins, while not as effective at water binding as sugars or minerals, can still make important contributions to retaining moisture, preventing freezing and microbe growth.

AGED MIMOLETTE CHEESE

Some water remains even after years of aging due to proteins with the help of sugars and minerals.

PROTEIN BAR

MILK POWDER

Powdered protein can be used to bind water and make food shelf stable.

PROTEINS / DISSOLVING

BEET SORBET
Egg white is added to prevent water from crystallizing and prevent grittiness.

BARBECUE PORK RIBS
Dissolved gelatin retains moisture in sticky glaze.

JERKY STICK
Dried meat is preserved because of the water-binding power of proteins in addition to sugars and minerals.

BROWNING

In the Sugar chapter, we talked about two kinds of browning: caramelization and Maillard browning. Caramelization comes from heating pure sugars until they explode into delicious fragments. This takes a lot of heat. Maillard browning adds proteins to the mix to help jump-start the process, and less heat is required.

Sugars are some of the most stable Ingredients. It takes a lot of heat to shatter them into the complex shards we love in caramelization. Proteins help generate complex flavors by loosening sugars up to explode and joining in the explosion themselves. Proteins are the fuses that light the sugar bombs. Once sugars start to decompose, proteins join the party, adding layers of flavor with scraps from *their* disintegration. Each type of protein has a slightly different combination of amino acid building blocks, each of which breaks apart into different shapes and sizes. The basic "starter set" of Maillard browning reactions gives us the generally roasty flavor of browned toast, coffee, meat, chocolate, and vegetables, while the compositional nuances of each type of food contribute to their different flavorful top notes.

The amino acids from which the proteins are made actually have much more to contribute to the Maillard browning process than the proteins themselves. Similar to the breakdown of carb chains to yield more reactive sugars (see the Carbs chapter), fragments of protein have more exposed, reactive "fuses" to ignite the browning process than intact proteins.

Regardless of the type of protein or amino acid, there is no set temperature at which Maillard browning officially starts. Foods like steak, bread, carrots, and mushrooms, however, need to hit temperatures above 100˚C in order to brown quickly enough to keep their interiors moist and/or juicy. The race to brown some foods before they dry out is why grills, ovens, and fryers are the places we most often associate with Maillard browning, but it can happen almost anywhere. Balsamic vinegar browns in a cellar. Egg powder and powdered milk brown in a refrigerator.

At any temperature, some of the proteins and sugars in food have enough energy to break into Maillard fragments. Food takes months to brown at refrigerator temperatures because Maillard browning is a probability game. Raising the temperature increases the number of molecules that are excited enough to explode. With enough proteins and sugars, the only limiting factor is how long we're willing to wait.

Fragments of protein have more exposed, reactive "fuses" to ignite the browning process than intact proteins.

Apart from cranking up the heat and adding more proteins, amino acids, and sugars, we can also change Maillard browning by fiddling with pH. Bringing the pH up (making food more alkaline) makes the protein fuse burn faster, so Maillard browning happens quicker. This is why we put lye on pretzels, and why adding baking soda can hold the key to making roasted vegetables and meat turn a deeper shade of brown.

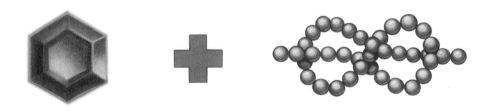

SUGARS + PROTEINS = MAILLARD BROWNING

PROTEINS / BROWNING

Proteins combine with sugars to make their own contributions to the complex explosion of taste, aroma, and color that arises from Maillard browning. Fragments of proteins and free amino acids brown more than whole proteins.

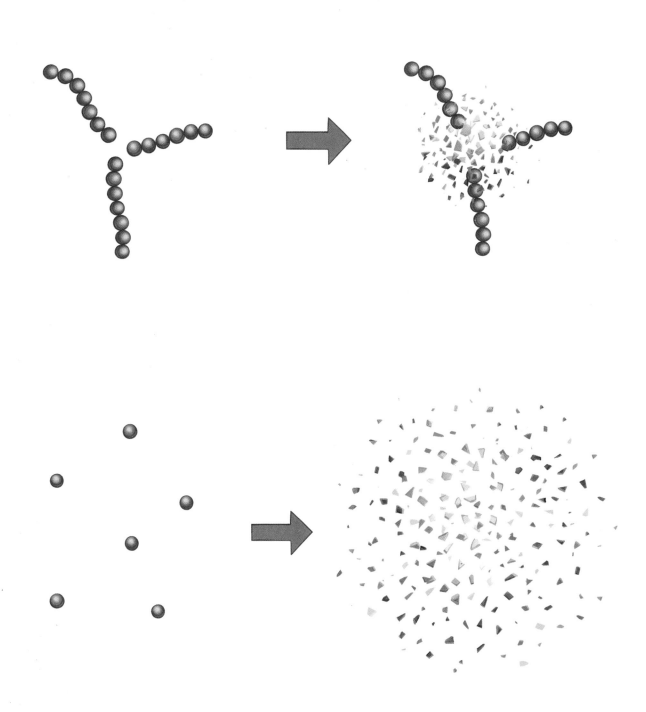

SMOKED TURKEY LEG

———

SEARED SCALLOPS
Proteins in meat contribute to
Maillard browning.

ROASTED ONION
Even vegetables like onions have
enough proteins to brown.

PROTEINS / BROWNING

PRETZEL BUN

———

COFFEE BEANS

These both have alkaline pH, which makes them brown quickly.

S'MORES

The trace amounts of proteins in chocolate brown during roasting of the cacao bean, the gluten in the graham crackers browns during baking, and the gelatin in the marshmallows browns during toasting on a fire.

EMULSIFIERS

When proteins unfold, the exposed water-hating parts of the chain start to panic and look for a place to avoid contact with water. In the case of coagulation, that salvation comes from sticking to another protein. If other proteins aren't within reach, a panicked, exposed protein will glob on to the first nonwater thing it finds. Desperate proteins will take any port in a storm, and the surface of a nearby lipid droplet is often the best sanctuary.

Just like lipid emulsifiers, proteins can coat the surface of lipid droplets. The proteins snake themselves in and out of the droplet's surface, orienting themselves so that both the water-loving and water-hating parts are happy. The coated droplets bounce off of one another rather than coalesce into one big oil slick. Proteins in an egg yolk help us get enough oil into a mayonnaise to make it creamy. Whisking the egg prior to adding the oil unfolds the proteins and primes them to latch on to each droplet. Heating the eggs before making a hollandaise unfolds them further, enhancing the effect. Salting and grinding the meat proteins to make sausages and pâtés also helps to keep lipids evenly distributed by littering the landscape inside the food with unfolded proteins. In the Lipids chapter, we discussed how lipids can act like spacers in gluten networks to make rich breads like brioche and challah more tender. The other perspective of that relationship is that the gluten proteins coat the lipid droplets and keep them evenly spaced from each other, allowing us to work a massive amount of tasty butter into a relatively small amount of dough.

Proteins flock to the surface of bubbles just as they flock to the surface of lipid droplets.

There is another type of safe haven for a water-hating protein in need: air. Gas bubbles are mostly empty space, and if you hate water, that space looks like an attractive alternative to watery discomfort. Proteins flock to the surface of bubbles just as they flock to the surface of lipid droplets. If there are enough proteins, they form a webby network that immobilizes the bubbles and creates a stable foam. This is one of the reasons why the bubbles in a cappuccino stick around for longer than the bubbles on top of a milk shake—steamed milk in the cappuccino has more unfolded proteins than the milk shake. Whipped meringues, airy cheesecakes and mousses, the giant holes in a loaf of ciabatta, and the bubbles on the outside of pork rinds are all due in part to proteins lending a hand in holding on to gas bubbles.

A side effect of all of this protein action is that taste and aroma get wrapped up in the proteins as well. Flailing protein chains act like sticky tentacles—anything they grab is dragged down into the depths. This can be a good thing, like when we want to capture smoke aroma in everything from barbecue to Gouda, but when it goes too far, it can kill the flavor of a dish. As with carb chains (see the Carbs chapter), taste or aroma compounds that get bound too tightly by proteins will stay in the food rather than escape to your tongue or into your nose at the right moment. If that moment passes, the food will already be in your stomach before the taste and aroma have had a chance to announce their presence. This means too much egg, dairy, soy, or any other protein can mute flavor, so make sure to begin with starting material flavorful enough to withstand getting knocked down a notch or two. If a blueberry puree, lemon juice, or spice mixture is concentrated enough, the resulting blueberry yogurt, lemon curd, and spicy sausage will deliver an appropriate flavor punch even after the protein has taken some of the taste and aroma out of the fight.

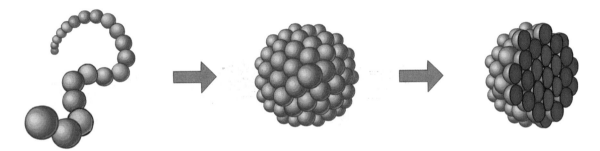

CROSS
SECTION

WATER-HATING AMINO ACID (GREEN), WATER-LOVING AMINO ACID (RED)

PROTEINS / EMULSIFIERS

Protein chains fold up in shapes that hide the water-hating amino acids on the inside and expose the water-loving amino acids on the outside.

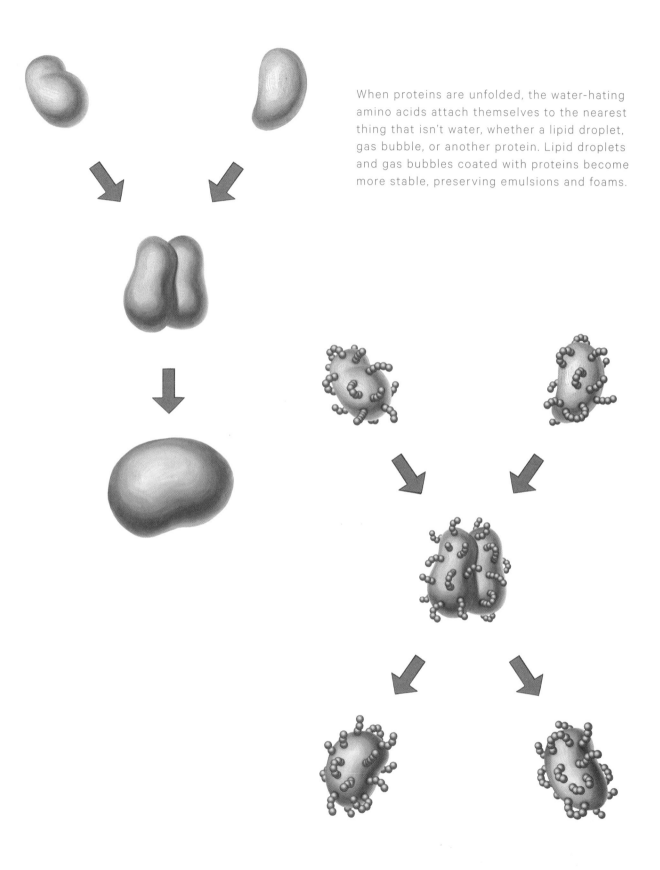

When proteins are unfolded, the water-hating amino acids attach themselves to the nearest thing that isn't water, whether a lipid droplet, gas bubble, or another protein. Lipid droplets and gas bubbles coated with proteins become more stable, preserving emulsions and foams.

FLAN

Egg and dairy proteins keep custard smooth and creamy.

BLACKBERRY YOGURT

Dairy proteins keep lipids and water suspended and hold on to the aroma of berries.

PISTACHIO ICE CREAM

Taste and aroma of pistachios are muted by dairy and egg proteins.

PROTEINS / EMULSIFIERS

SMOKED GOUDA
Proteins and lipids trap smoky aromas.

LEMON CURD
Taste and aroma of lemon are muted by egg protein, which also helps preserve butter emulsion.

COCONUT MERINGUE
Egg proteins mute the release of coconut aroma and preserve airy structure.

HERBED SAUSAGE
Meat proteins hold on to herb aroma and preserve emulsion.

ENZYMES

A lot of proteins function as glorified scaffolding—they stand still and provide support. Some proteins, however, have the ability to come alive and carry out the complex tasks that make life possible. These protein mini-robots are called enzymes, and they are useful tools for a cook who knows what they are and how to use them.

Enzymes work by bringing together the microscopic *mise en place* for Ingredient reactions and creating a good environment in which those reactions can happen. Enzymes are the ultimate molecule matchmakers. In our cells, and in the cells of every living thing, enzymes are responsible for building things up and tearing them back down. These are the same basic functions that the most useful enzymes can play in our food—they're either "builders" or "breakers."

The "breaker" enzymes work like microscopic scalpels, giving us the ability to selectively slice and dice Ingredients from the inside to create amazing textures, tastes, and aromas. These types of enzyme can chop proteins and carbs into individual amino acids and sugars, which season everything from aged meat, cheese, and soy sauce to malted barley, bread, and sweet potatoes with natural savoriness and sweetness. They also break down lipids, which can create either pleasant or overwhelming aromas, to which anyone who has ever tasted goat cheese can attest. Apart from taste and aroma, breaking down large Ingredients like carbs and proteins has profound effects on texture. Aged meat and long-fermented bread dough become more tender as tough muscle fibers and gluten are clipped to pieces, cheeses turn to goo as their dairy protein skeleton is digested, and hard, starchy grains and legumes like rice and soybeans turn to spreadable pastes over time.

> Whether they build things up or break them down, each enzyme can process only one type of Ingredient.

The "builder" enzymes have a less impressive set of applications, and we usually try to discourage their activity in the kitchen. They turn individual sugars in English peas into starch after they are harvested, use gases like oxygen to build brown pigments in apples, avocados, tea leaves, and potatoes, and create sticky, thick bundles of carbs in fermented foods like kombucha, vinegar, and natto. While they don't attract a lot of attention in the kitchen, the food industry uses builder enzymes to manufacture big molecules in massive quantities, which is where we get things like xanthan gum and some guilt-free protein alternatives to animal meat.

Whether they build things up or break them down, each enzyme can process only one type of Ingredient. In some foods, the specific enzymes we need are naturally lurking within, waiting to do our bidding. Other foods lack the enzymes we're looking for. With those foods, we can either mix them with enzyme-rich ingredients or create enzymes from scratch with the help of microbes. Tropical fruits, unpasteurized dairy products, and miso are abundant sources of protein-digesting enzymes, and egg yolks, sweet potatoes, and malted grains all contain lots of enzymes for breaking down carbs. These enzyme sources can be added to everything from marinades, pastes, and brines to batters, doughs, soups, and sauces.

The key to cooking with enzymes is knowing how to manage them. Sometimes, we want to encourage enzymes in garlic and onions to create a lot of pungent aroma, so we smash garlic cloves and mince onions finely to release as many enzymes as possible from their cellular cages. Other times, we want to use garlic and onions as a background foundation for another flavor, so we need to minimize the work done by those enzymes. Luckily, all enzymes are proteins and thus play by protein rules. This means that if we want an enzyme to quit doing its thing, we need to make it uncomfortable.

Uncomfortable enzymes unfold, and unfolded enzymes don't operate. The strategies for unfolding enzymes are the same as with any protein: heat them up, add a lot of minerals or sugar, change the pH, or physically beat the heck out of them. In the case of garlic and onions, we can blanch, salt, or pickle them before peeling or cutting them, which allows us to deactivate the enzymes

If we want to make enzymes work as much as possible, we need to make them comfortable.

before they have a chance to get out of their cages and burn our noses with aroma. If we want to make enzymes work as much as possible, we need to make them comfortable. Brewing, wine making, aging cheese, and making miso are crafts that have been refined over centuries. The artisans who make those foods have discovered the perfect combination of temperature, time, pH, and other variables to make enzymes and the microbes that wield them as comfortable as possible to achieve stellar results. Now that we understand those patterns, however, there is no need to stick to the path of tradition. Knowing a particular enzyme's turn-ons and turn-offs allows us to bend it to our culinary will. We can apply the same enzyme-farming techniques that we once used on grapes, milk, and grains to other foods—virtually anything with protein can become a savory explosion of taste, and anything with starch can become sweeter with no added sugar.

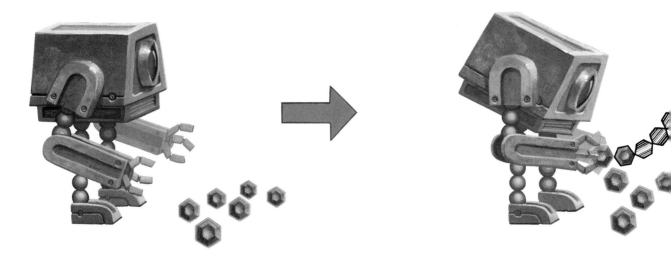

"BUILDER" ENZYME

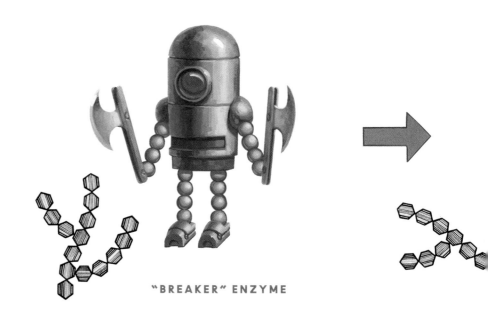

"BREAKER" ENZYME

PROTEINS / ENZYMES

The most important enzymes in the kitchen either build
molecules up or tear them apart.

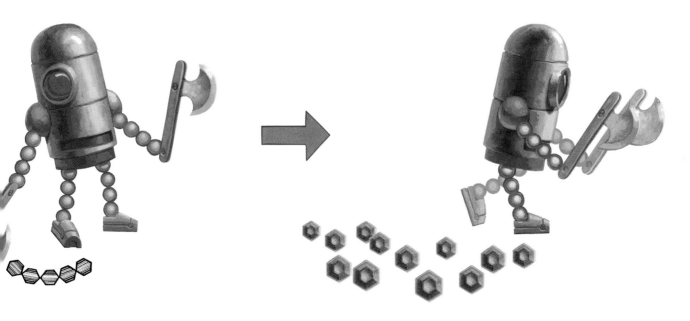

CAMEMBERT
Protein-digesting enzymes make cheese gooey and soft as it ripens.

GARLIC CHIVES
Aromas in the onion family are produced by enzymes that activate when tissues are crushed or bruised.

BASIL
Enzymes activate when leaves are damaged and turn them brown when exposed to oxygen.

PINEAPPLE
Tropical fruits contain enzymes that digest proteins; all fruits contain enzymes that produce aroma.

PROTEINS / ENZYMES

PAPAYA

Tropical fruits contain enzymes that digest proteins; all fruits contain enzymes that produce aroma.

BLACK TEA LEAVES

Same browning enzymes as in basil, potatoes, artichokes, and apples are activated by crushing the leaves to give tea its color.

BREAKDOWN

Cooking with proteins is all about managing the chain, and that includes breaking the chain apart into individual links. Dismantling proteins into individual amino acids completely changes the way they behave.

Breaking the chain affects how proteins link up to form structures. Amino acids can't make gels. This is why creamy cheeses like Brie become creamier as they ripen. Microbes in the cheese make enzymes to chop the proteins into bite-size pieces. Protein breakdown causes the scaffolding of the cheese to soften, oozing creamy lipids and savory amino acids all over the place. This same mechanism is why firm tofu becomes spreadable over time when fermented.

Intact proteins don't really taste like anything. Breaking proteins into smaller pieces releases a rainbow of tastes, from sweet and sour to bitter, salty, and, most prevalent, umami. Amino acids also bind water better than intact protein, just like sugars bind water better than intact carbs. Oysters and other sea creatures use water-binding amino acids as a way to survive. The ocean is full of minerals, and those salts jealously hoard water. Oysters don't have waterproof skin, so they stockpile amino acids along with sugars and minerals of their own, which bind water and prevent the ocean salt from drying them out. This Ingredient arms race between oysters and the ocean works out well for us because those stockpiled amino acids, sugars, and minerals season the oysters from the inside out.

Breaking proteins into smaller pieces releases a rainbow of tastes, from sweet and sour to bittery, salty, and most prevalent, umami.

The most dramatic change after protein breakdown comes from Maillard browning. Amino acids are the active ingredient in the protein fuse that catalyzes the reaction, so the more aminos that are exposed the better the reaction. Long-aged meat, bread, cheese, and condiments brown much more evenly than food with intact proteins. Marinating food with ingredients that have a lot of broken proteins causes browning to go wild. Adding soy sauce, cheese, fish sauce, or any other food with lots of free amino acids can even cause food to brown too quickly, burning on the outside before it can cook through.

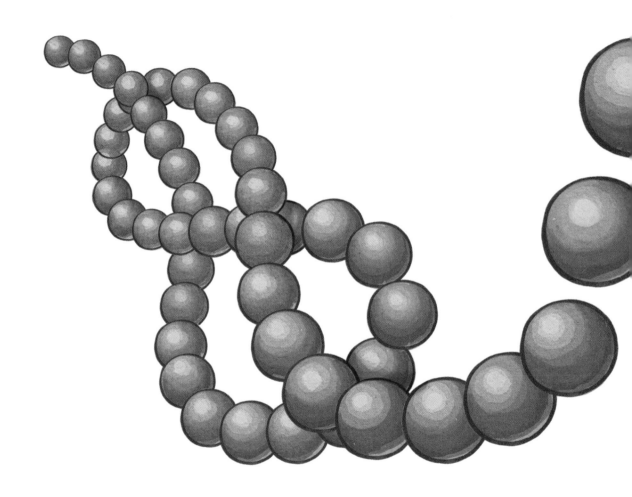

PROTEINS / BREAKDOWN

Broken bits of proteins make food tender and savory,
bind water better, and brown faster.

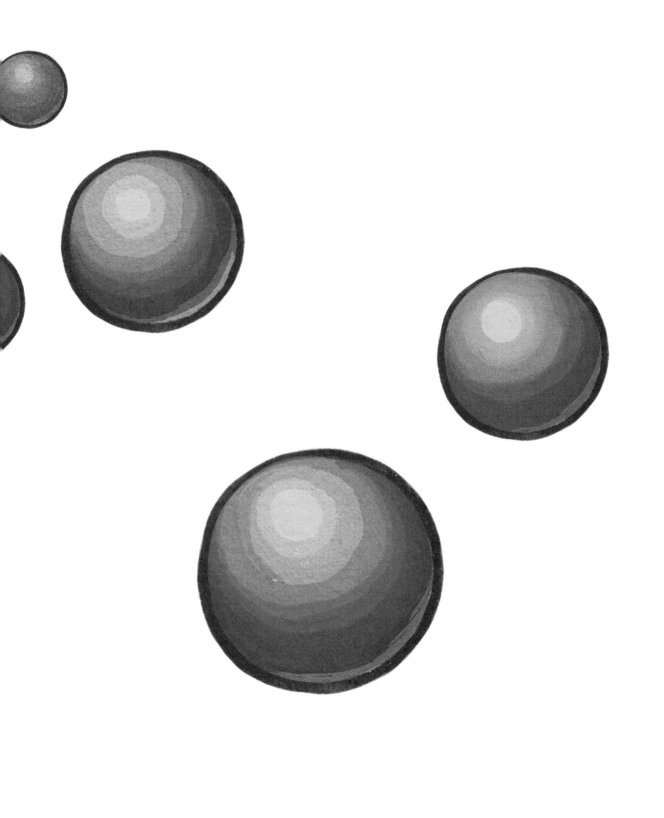

DRY-AGED BEEF RIBEYE

Protein breakdown by enzymes
in meat increases savoriness,
tenderness, and browning.

PROTEINS / BREAKDOWN

DORITOS

Being seasoned with broken-down protein from several different sources makes them the most savory thing ever.

SOY SAUCE

Soy protein breakdown during fermentation creates savoriness and increased browning.

KATSUOBUSHI (BONITO FLAKES)

Broken-down fish protein forms a savory base for soups and sauces.

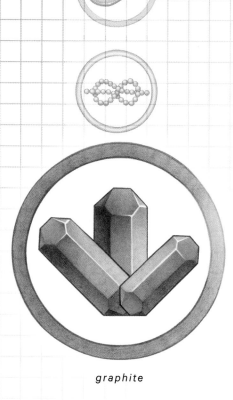

graphite

MIN

ERALS

Minerals are rocks. We eat a lot of rocks, and they do a lot more for us than make stuff salty. Minerals affect everything from the crispness of fermented vegetables to the color of fresh meat. Although minerals are found in much lower concentrations in food than any of the other Ingredients, they pack a serious punch and can affect food in four ways:

- They have **TASTES**.
- They **BIND BIG MOLECULES** together.
- They **DISSOLVE AND CRYSTALLIZE**.
- They make **COLORS**.

TASTE

Table salt—sodium chloride—is salty, obviously. It's the most important mineral for the taste of food, but it's not the only one that lights up your tongue.

When a mineral enters your mouth, taste buds grab it. They feel it out and tell the brain what they notice. Iron and copper taste gamey, organ-y, and somewhat like blood. Like all minerals, iron and copper are everywhere, but they're especially concentrated in organ meats like liver and heart, the bloodline of fish fillets, wild game animals, and all dark meat. Magnesium and calcium taste saline, slightly bitter, and marine, depending on the other minerals that accompany them. They flavor gray sea salt, tofu, seaweed, shellfish, and a bunch of other things. Potassium tastes like bitter metal. Somebody once thought that substituting potassium for sodium in low-sodium hot dogs was a good idea. It was not. Apart from table salt, we only enjoy the taste of minerals at very low levels.

We can't smell minerals—at all. To smell something, it has to be able to float into your nose to reach your smell receptors. Minerals don't float. In spite of the fact that we can't smell minerals, ocean air smells salty and blood smells iron-y. Those smells come from our memories. When we smell ocean air, we smell decayed bits of lipids and other Ingredients from kelp, sea urchins, and other smelly things that live

Our minds use those memories to season aromas with echoes of taste.

and die in the water. Our brains associate that cocktail of smell with the taste of salty seawater, so we think, "Ah, the salty smell of the sea." The same type of thing happens with blood. We can't smell iron, but our brains can recall the taste of gamey meat or the bloody childhood experience of losing a tooth. Our minds use those memories to season aromas with echoes of taste.

In addition to the psychological effects minerals spark in our memories, minerals can also make indirect physical contributions to aroma. Minerals have the ability to react with other Ingredients, especially lipids, and those interactions can cause lipids to break apart into smelly pieces, as is the case with iron generating some of the aromas of leftover, reheated food.

MINERALS / TASTE

Different minerals evoke different tastes, including the slightly bitter magnesium in tofu, the gamey iron of red meat, and the familiar saltiness of sodium chloride on chips.

TOFU

TASTE BUD

HAMBURGER

POTATO CHIP

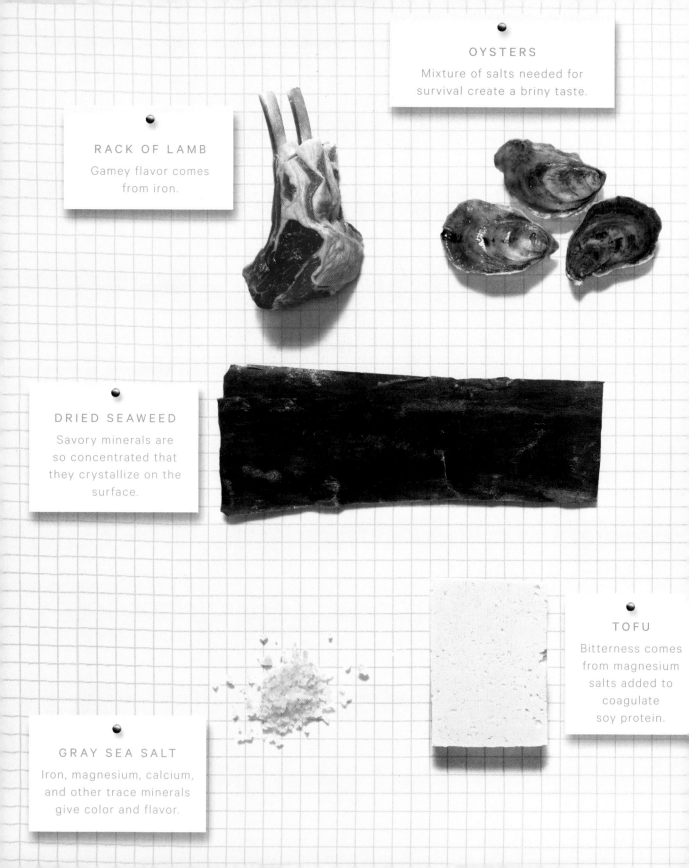

OYSTERS
Mixture of salts needed for survival create a briny taste.

RACK OF LAMB
Gamey flavor comes from iron.

DRIED SEAWEED
Savory minerals are so concentrated that they crystallize on the surface.

TOFU
Bitterness comes from magnesium salts added to coagulate soy protein.

GRAY SEA SALT
Iron, magnesium, calcium, and other trace minerals give color and flavor.

WHITE SEA SALT
Pure sodium chloride

HOT DOG
Low-sodium versions use potassium chloride, which tastes like bitter metal, instead of table salt.

TILAPIA FILLET
Iron in bloodline creates gamey taste and generates fishy aroma by causing lipid breakdown.

MINERALS / TASTE

BINDING BIG MOLECULES

Carbs and proteins are the largest Ingredients, which is why they affect texture more than the other Ingredients. They act like steel beams to reinforce the structure of food and form cages that trap water. Some minerals work like rivets to weld those structures together.

Minerals latch on to the surface of carbs and proteins, drawn there like barnacles to the hull of a ship. All minerals stick to these giant Ingredients, but some can grab on to more than one chain at a time. Calcium, magnesium, and aluminum each have two or more "arms." With those arms, they grab on to two strands of a chain and cinch them together. If there are enough of these links, they can knit big molecules into the netlike shapes required to trap water as a gel. A lot of traditional and modernist food gels are set by this same basic mechanism—everything from tofu, low-sugar jam, and ricotta to spherified alginate pearls, eggless custards, and heat-stable ice creams. If a gel has already formed, adding minerals can help strengthen it. Canned whole tomatoes are treated with calcium to preserve their shape. The calcium strengthens the natural pectin gel in the tomato tissue, helping it withstand the inferno that is an industrial canning machine.

A small dose of minerals can help larger Ingredients transform the texture of food, but they don't do much on their own. Since minerals are so tiny, they don't present a meaningful roadblock to water. The amount of salt required to noticeably thicken food by itself would make any food inedible, so we use minerals behind the scenes for texture rather than feature them as the star.

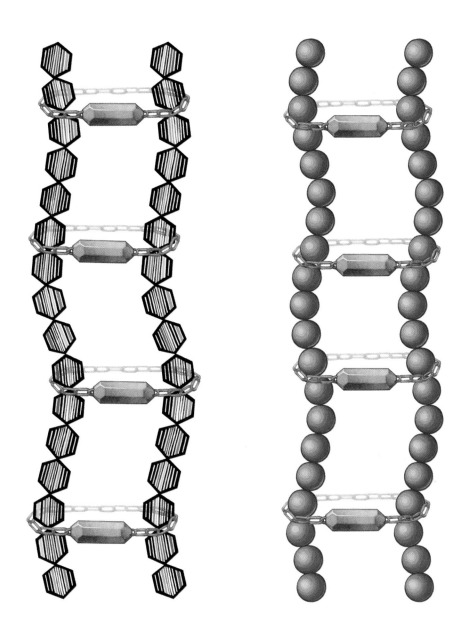

MINERALS / BINDING BIG MOLECULES

Minerals like calcium, magnesium, copper, and aluminum can cinch carbs and proteins together to create and strengthen gels.

PICKLED CUCUMBER
Calcium added to keep crispy after pickling; aluminum was traditionally used but abandoned because of health concerns.

PÂTÉ DE FRUITS
Calcium is added along with pectin to cause gel to set.

ASPARAGUS
Calcium in carb scaffolding dissolves during cooking, causing vegetables to soften.

MINERALS / BINDING BIG MOLECULES

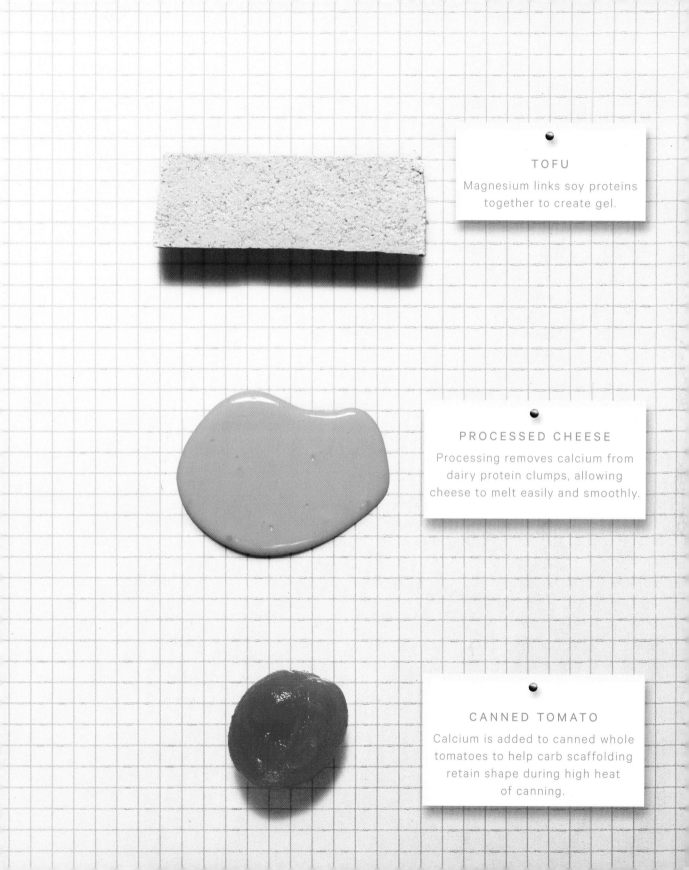

TOFU

Magnesium links soy proteins together to create gel.

PROCESSED CHEESE

Processing removes calcium from dairy protein clumps, allowing cheese to melt easily and smoothly.

CANNED TOMATO

Calcium is added to canned whole tomatoes to help carb scaffolding retain shape during high heat of canning.

DISSOLVING

Minerals may not present much of a physical roadblock to water, but along with sugar, they can hold its attention better than the other Ingredients. Water breaks mineral crystals apart into individual molecules. Each mineral molecule receives a captive audience of water molecules, which in turn have a harder time freezing, evaporating, and helping microbes grow.

Minerals have a hypnotizing effect on water that keeps water molecules from organizing themselves into ice crystals. We scatter salt on sidewalks in the winter to keep ice from forming, and the same principle applies to food. Delicate foods like frozen shrimp, fish, and chicken tenders are often pumped full of salts to preserve their texture. The dissolved minerals prevent water from forming large ice crystals that poke, tear, and shred the fragile scaffolding of frozen food.

If we want to remove water from food to preserve it or change its texture, the most effective approach is to combine salting with drying. When we add salt or other minerals to solid food, water travels to the surface, dissolving the minerals to create a concentrated, salty liquid. At this point, we can wipe the salty moisture off the food, hang the food to allow air to circulate, and/or heat the food to dry it quickly. The salt gives us a jump-start on the overall drying process by bringing water to the surface of the food where it can be whisked away with ease, which is why everything from cured salmon and jerky to cured egg yolks and bottarga starts with a thorough salting.

If we leave the food in that salty liquid for too long, however, the process can drive water in the opposite direction. As more and more water comes to the surface, the salt becomes more and more diluted. Eventually, the concentration of salt on the inside and outside of the food becomes about the same. This gives the salt on the inside of the food a fighting chance at attracting the water that it lost, and the water begins to march back to the interior. This phenomenon may make it harder to dry food properly, but we can also leverage it to our benefit to yield incredibly juicy food.

Minerals may not present much of a physical roadblock to water, but along with sugar, they can hold its attention better than the other Ingredients.

Once salty water penetrates a turkey, a pork chop, or a piece of cod, it tends to linger there, keeping the meat from drying out as quickly during cooking and hot holding. The charismatic pull of dissolved minerals keeps water from getting excited enough to fly away and evaporate. This approach is commonly used for events, buffets, and large family dinners where food sits around for extended periods of time, but we can apply it to any situation where food needs a juicy boost. While dry salting the surface will get us there eventually, using a brine can be a much more straightforward process. By pre-dissolving the minerals (and sometimes sugars) in water before applying to food, brines help us skip to the moment where diluted salty water is drawn into the interior of the food. A word of caution: the extra water taken up by the meat may make it juicy, but that water also dilutes the taste and aroma of the meat. When cooking with bland, factory-farmed meat, poultry, and seafood, brining can represent a choice between juicy texture or flavorful taste and aroma.

High levels of minerals like salt prevent microbes from being about to grow on our food.

Marooned sailors knew that salty seawater was undrinkable—"Water, water, everywhere, / Nor any drop to drink." If microbes could read, they'd probably agree with Coleridge on that one. High levels of minerals like salt prevent microbes from being able to grow on our food. This is the key to salted capers, olives, lemons, fish, meat, cheese, and anything else we want to preserve. Any food can be made immortal with the sufficient addition of minerals; the only question is whether it tastes good salty or not.

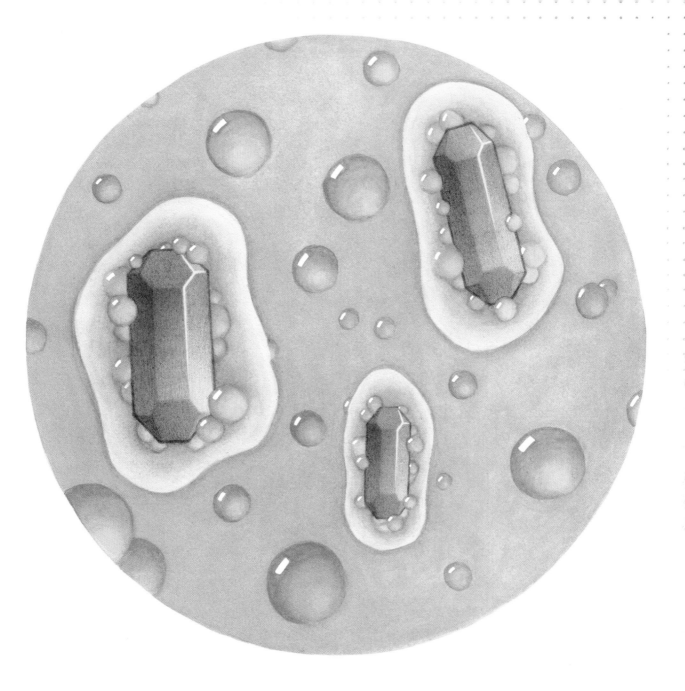

MINERALS / DISSOLVING

Minerals, along with sugars, are the most effective Ingredients when it comes to water binding. We use them to control evaporation from food during cooking, protect food from ice crystal damage during freezing, and discourage microbial growth.

CAVIAR

Salt added to preserve eggs also retains water inside, helping them stay plump.

PRESERVED LEMONS

———

UMEBOSHI (SALTED PLUMS)

———

CAPERBERRIES

Salting fruits favors the growth of lactic acid bacteria (which tolerate more salt) over yeast, turning them into tangy pickles rather than booze.

PANCETTA

Salting at the beginning of the curing process wards off enough microbes to give the meat time to age, dry, and mature.

CURED ANCHOVY

Salt helps the flesh stay firm during canning by retaining moisture in the meat.

MINERALS / DISSOLVING

COLORS

When light bends and shifts, we see color. Anything that bends light the right way will create colors. In food, color can come from scraps of lipids, sugars, and other Ingredients, but minerals give us one of the broadest palettes of colors to paint with.

Two of the most common colors in food are green and red; both are made possible by minerals. Iron and magnesium sit in the middle of two types of pigment molecules: heme (found in hemoglobin and myoglobin) in meat and chlorophyll in plants. These pigment molecules are shaped like solar panels, precisely tuned to interact with light. The mineral sits at the center, anchoring the array and holding everything together. Any jostling of the mineral will change the shape of the pigment, which changes its color.

Acid can kick magnesium out of chlorophyll, which turns the chlorophyll brown. This can happen slowly, like when a salad gets dressed too early, or quickly, as broccoli simmers in a pot of tap water, which is acidic in a lot of places. Acid from the cooking medium isn't the only danger—there are natural acids inside plants that move around and wreak havoc during cooking. This is why adding baking soda keeps things super green—baking soda neutralizes any acid that may be lurking, whether in the pot or in the broccoli itself.

Any jostling of the mineral will change the shape of the pigment, which changes its color.

Keeping cooking times to a minimum also helps preserve the green color, since magnesium is more likely to slip out of the chlorophyll when it gets hot and excited.

Acid and heating can change the color of red meat as well, but much more quickly. Not only is the iron vulnerable, but also the pigment built around it is made of protein, which has its own sensitivities (see the Proteins chapter). This is why marinated meat turns brown on the surface and why meat goes through a spectrum of changes, from red to brown, as it cooks. Iron even calls the shots in meat color before cooking ever starts. Iron exists in blood to serve as a vehicle for carrying oxygen around, and the color of the pigment acts like a signal for how much oxygen there is. Meat wrapped in plastic or packed in a vacuum-sealed bag can become dull and gray. Oxygen in the air is why meat blooms back to rosy life after sitting on an open counter. Some butcher shops flood their display cases with oxygen to enhance this same effect. There are entire industries dedicated to preserving the color of meat and vegetables, and most of those techniques revolve around these same principles—keeping the minerals in those pigments happy and where they should be.

When all else fails, make something new. Artificial coloring agents take more inspiration from the rocks that bore the minerals in the first place than from anything else. Most food colorings don't have the same mineral–solar panel shape as chlorophyll or hemoglobin. They are molecules that come from oil refineries,

Artificial coloring agents take more inspiration from the rocks that bore the minerals in the first place than from anything else.

plants, animals, microbes, and the earth, but a connection they share is their stabilization by minerals. A lot of dyes are reactive, packing a colorful punch on whatever they first come into contact with. That makes them difficult to transport. Because of this, many dyes are paired with sodium, calcium, or other minerals to bind them and make them a bit less reactive. Minerals can work like the safety on a color gun. Once it's time to put these colors into candy corn, ice pops, and bright green mint chocolate chip ice cream, manufacturers follow a series of protocols to dissolve the minerals, unlocking the colors that the minerals protect.

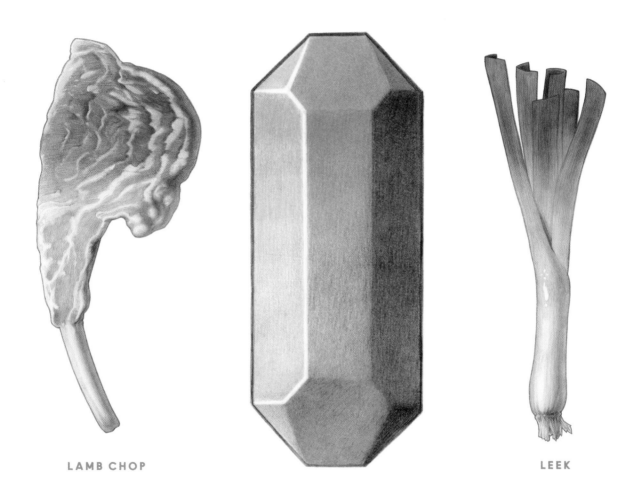

LAMB CHOP

LEEK

MINERALS / COLORS

Iron is responsible for the red color in meat, and magnesium is integral to the green color of plant foods.

BEEF BRISKET

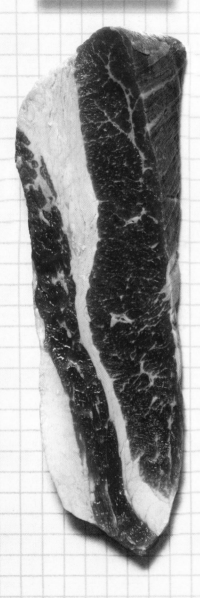

PINK SALT

TUNA LOIN

Iron is responsible for
the red color in meat and
unrefined salt.

MINERALS / COLORS

Magnesium plays a crucial role in making chlorophyll green in all plant foods.

MINT

COLLARD GREENS

CUCUMBER

charcoal

ES

Gases (which technically include gaseous water) are the flightiest Ingredients. They're agile, so the most important trick to working with gases is herding them from place to place. Keeping gases out is sometimes just as important as keeping gases in. Whether we're trapping bubbles in a foam or protecting an avocado from brown-inducing oxygen, we use gases for four basic reasons:

- They **DISSOLVE**.
- They **CREATE BUBBLES**.
- They **CAUSE REACTIONS**.
- They **EXPAND AND CONTRACT**.

DISSOLVING

Gases can dissolve in water, but they do it on their terms. Dissolving gases deviates from the normal Ingredient routine because of how light they are. Water can surround each gas molecule with an entourage of water molecules, but it has a hard time holding on to them. When dissolving gases, we need to do a few things differently to keep them from flying away.

Gases like to dissolve when it's cold. This is the opposite of the other Ingredients, which generally dissolve more easily with heat. When dissolving solids and liquids, adding heat allows the water to cover more ground, guarding more dissolved bits with the same number of water molecules. Gases, however, also move faster with more heat, much faster than water on its best day. When they get hot, gases slip past the water mob easily. Keeping the temperature down helps keep gases where they are, which is why we chill beer, soda, champagne, and other fizzy drinks.

Gases are nimble, so they'll escape eventually, regardless of temperature. To keep gas dissolved indefinitely, we need pressure. Gases want somewhere to go once they leave water, and pressure deprives them of the space they need to spread out. Putting drinks, whipped cream, Cheez Whiz, and anything else in cans or bottles is our most effective way of making sure that gases stay where they are. Cans and bottles are designed to allow gases to move around, but only when we say so. They jump and fizz on our tongues, but only when we open the

> Dissolving gases deviates from the normal Ingredient routine because of how light they are.

container. Anyone who has ever used a vacuum sealer has seen the other side of this concept. **Pressure pushes gases down to dissolve, but a vacuum makes it easier for them to fly out of solution.** Room-temperature soups, sauces, and anything else with dissolved gas appear to burst into boil as the vacuum sucks away the atmospheric pressure. Even at room temperature, dissolved gas is raring to escape and bubble out of solution. Turning on a vacuum pump pulls off the invisible atmospheric weight pushing down on the dissolved gas and gives it the opportunity it needs to make a getaway.

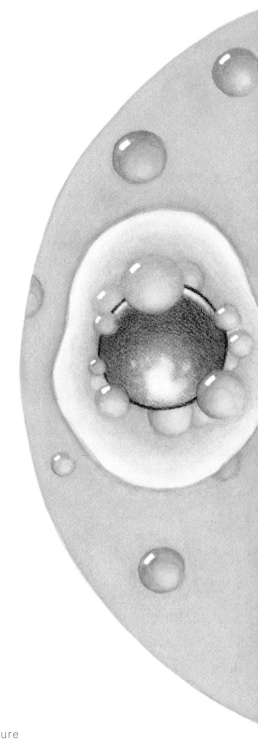

GASES / DISSOLVING

Gases play by different rules for dissolving than the other Ingredients. Adding heat makes them dissolve less, and pressure is our greatest tool for keeping them dissolved.

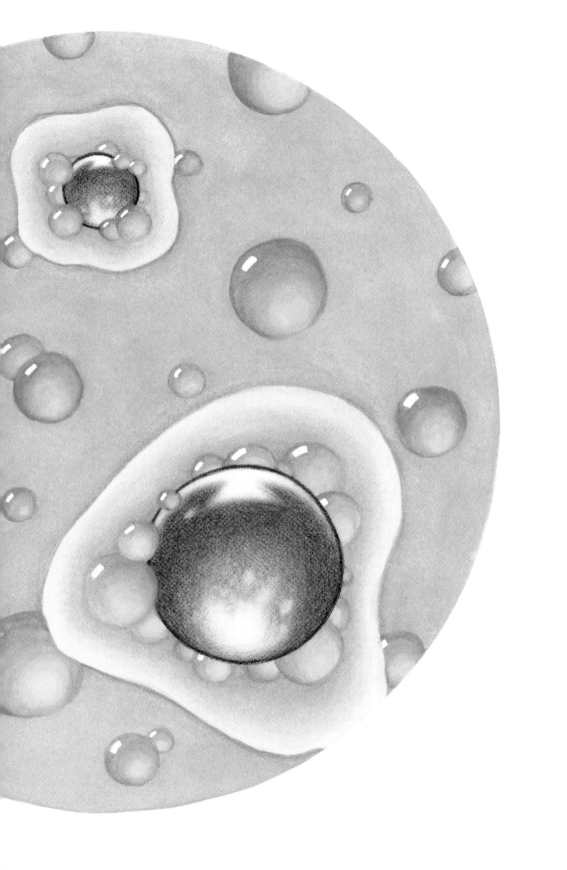

CHAMPAGNE

Dissolved carbon dioxide makes these foods taste acidic and sour. When the pressure from the bottle, can, or fermentation crock is opened, the escaping gas fizzes on the tongue.

GASES / DISSOLVING

COLA

KIMCHI

CANNED WHIPPED
CREAM
(Acidic, fizzy whipped cream
would be weird, so we use
flavorless nitrous oxide instead.)

BUBBLES

When gases come out of solution, they can make bubbles. Bubbles affect the texture, appearance, taste, and aroma of food—they're the gas equivalent of emulsions.

When gases slip the bonds of their water captors, they escape by flying out to the open air. On the way up through the water, they meet other gas molecules, forming bubbles. Bubbles form fastest when the gases have somewhere to congregate. They are born on imperfections in a container, bits of undissolved solids, the tines of a whisk, or even other bubbles. This is how the finest champagne flutes achieve a single stream of bubbles: The inside is perfectly smooth except for a dimple at the bottom of the glass where bubbles can form. Once a bubble is made, filling it is easy. From there, we add heat and let the bubbles expand for maximum puffing effect. This is why we fold whipped egg whites into soufflés and allow breads to rise before putting them in the oven.

Dissolved in water, gas molecules do almost nothing to thicken liquids. It's too easy for water to dodge them. Bubbles, however, are excellent roadblocks for water. Just like the other Ingredients, bubbles have more thickening power when they are evenly dispersed. Fine bubbles on top of a Guinness make a thicker foam than the coarse bubbles on top of a Coke. Whipped cream siphons are designed to make small bubbles to give the most velvety texture possible. Whisking tempura batter not only makes for an airier crust on fried food, but the bubbles thicken the batter, helping it cling to the food for a more even coating.

Bubbles thicken liquids, but only when they stick around long enough to make a difference. Foams are what happen when gas bubbles get trapped in nets made of other Ingredients. These ingredients are usually proteins and carbs, which trap gas bubbles like buoys in a net. Even though sugars can't form nets, they help foams live longer by thickening the water between each bubble, making it harder for the bubbles to float to the surface and pop. Trapped gas bubbles are why everything from sauces to pasta water can boil over when cooking—as water evaporates, the concentrated Ingredients trap more and more bubbles, filling up the pot.

Another trick to keeping gas bubbles around is to use some of the same Ingredients that work as emulsifiers. Gas bubbles are mostly empty air, and anything that hates water will often choose empty nothingness over contact with water. This causes Ingredients like unfolded proteins and some lipids to flock to the surface of bubbles, coating them and preventing small bubbles from blobbing together, getting larger, and floating away even faster. We can find these emulsifiers in a variety of foods, from grains, legumes, and root vegetables to meat, dairy, and eggs.

Bubbles affect the texture, appearance, taste, and aroma of food—they're the gas equivalent of emulsions.

Bubbles are mostly empty space, and empty space has no color. In all plants, there are natural bubbles in the spaces between cells. Blanching green vegetables and using the modernist technique of pulling a vacuum on fruits like watermelon pulls the bubbles out of those spaces. When we remove those bubbles, we turn up the saturation on those colors. The same holds true for frothy liquids, which is why chocolate mousse and cherry sorbet are so much lighter in color than the chocolate sauce and cherry syrup from which they were made.

Since bubbles dilute flavor, foams can be bland.

Empty space also has neither aroma nor taste. Since bubbles dilute flavor, foams can be bland. Over-churned industrial ice cream and poorly prepared foamy sauces are two examples of how foams can be an annoying barrier between you and flavor. On the other hand, if executed properly, foams can be awesome. Aromas can erupt quickly from bubbles once they burst, so if the base liquid is concentrated enough, foamy food can knock you over the head with flavor in addition to delicate texture.

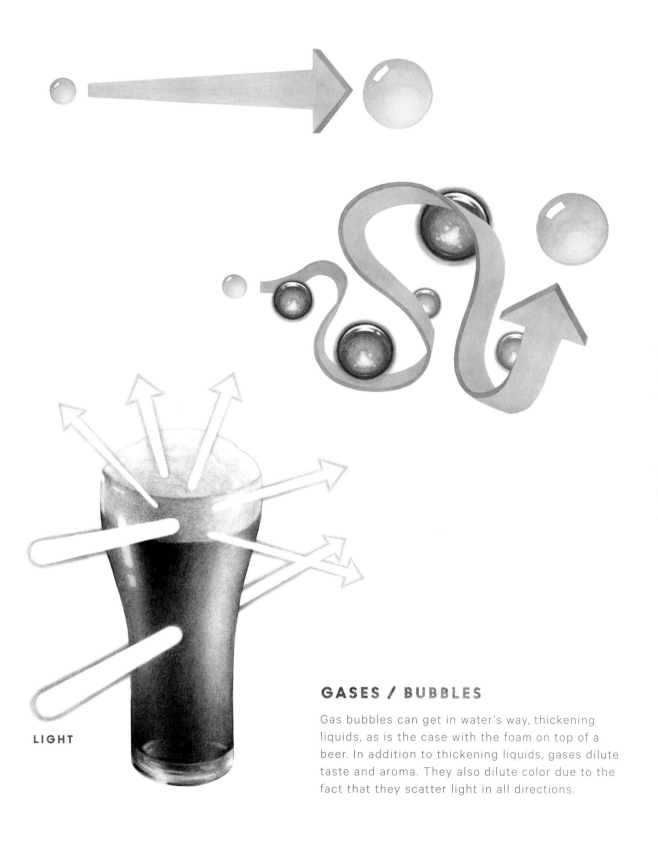

LIGHT

GASES / BUBBLES

Gas bubbles can get in water's way, thickening liquids, as is the case with the foam on top of a beer. In addition to thickening liquids, gases dilute taste and aroma. They also dilute color due to the fact that they scatter light in all directions.

MARSHMALLOWS

Marshmallows start out as a clear, yellowish sugar syrup, but the incorporation of tiny air bubbles turns them stark white.

CARAMEL HONEYCOMB

The bubbles formed from the caramel (acid) and baking soda (alkaline) neutralization create an airy texture when the candy hardens.

COMPRESSED WATERMELON

———

BLANCHED BRUSSELS SPROUTS

Removing air bubbles from food, whether with a vacuum pump or by heating food so the bubbles can escape, causes the color to intensify.

GASES / BUBBLES

RASPBERRY FOAM

—

CHOCOLATE MILK FOAM

Adding bubbles makes liquids become thicker and more luxurious, but it also dilutes taste, aroma, and color with flavorless air.

CHEMICAL REACTIONS

Gases wander around more freely than any other Ingredient. Like roving bands of outlaws, gases can descend on any food without warning. Keeping food safe from those raiders or offering it up to their whims is how we leverage gases for deliciousness.

The biggest culprit is oxygen. Silently and without any fanfare, it sneaks into food, wreaks havoc, and slips away. Oxygen is an assassin. If left to spiral out of control, these reactions cause food to deteriorate. Oxygen can break Ingredients down into pieces that smell horrendous, kill aromas that we want to preserve, turn bright colors dull, and provide fuel for enzymes to turn avocados, potatoes, artichokes, and apples brown. If properly controlled, this power can be harnessed to create some cool effects in food. Most tea leaves are crushed and allowed to brown in open air to deepen their color. We also enjoy what these reactions do to the aromas of fried foods, the pungent musk of melons, or the delicious funk of aged fatty meats like chicken confit and prosciutto.

Carbon dioxide has a more straightforward effect on food—it can make things acidic. When carbon dioxide comes into contact with food, it dissolves and lowers the pH. This is where we get the prickly, sour experiences of drinking soda, eating vigorously fermented kimchi, and cracking open Pop Rocks.

Oxygen and carbon dioxide are two of the most common gases produced by living things, but plants use another kind of gas to communicate with each other. Since

> When carbon dioxide comes into contact with food, it dissolves and lowers the pH.

they can't speak or move, plants often communicate with airborne hormones such as ethylene. These gases send messages like "RIPEN" or "GROW FASTER." We don't deal with them much in the kitchen, apart from when we store food together. Some fruits, like apples and bananas, send out these signals, and they can be picked up by anything that is nearby. These signals can make other fruit go from underripe to perfect to mush in a matter of hours, so be wary when leaving plants alone to congregate and conspire.

There are a lot of gases that we use specifically because they don't cause any trouble. Nitrous oxide, argon, helium, and other inert gases give us the chance to leverage the other gas functions without worrying about changing the food. Whipped cream and nitro-finished beverages allow us to enjoy bubbliness without the prickly sourness of carbon dioxide. Oxygen can ruin a bottle of wine once opened, so a lot of wine-saving devices flush oxygen out and replace it with an inert gas. Bagged salads come sealed with their own atmosphere of precisely formulated gases to keep spinach and radicchio from metabolizing themselves into a brown puddle.

GASES / BUBBLES

Reactive gases like oxygen have the power to change our food wherever they come into contact.

SEA URCHIN

Marine lipids are so vulnerable that as soon as they are exposed to oxygen, the smelly clock starts ticking.

OLIVE OIL

Exposure to oxygen makes delicate oils turn rancid, smelling like musty cardboard.

ARTICHOKE

One of the many foods, along with mushrooms and avocados, that have browning enzymes, artichokes turn brown when cut open, liberating enzymes and exposing them to the oxygen they need to change color.

GASES / CHEMICAL REACTIONS

GRAPES

Antioxidants like sulfites are added to wine to prevent it from oxidizing during storage.

FRENCH FRIES

The aroma of fried food comes from lipids being broken apart by oxygen.

ROASTED RED PEPPER

Combustion is fueled by oxygen.

EXPANSION + CONTRACTION

Gases expand and contract more aggressively than any other substance in the kitchen. A volcano of expanding gas is a powerful force for changing the texture of food. We talked about steam (gaseous water) in the Water chapter, but we can manipulate any type of gas to achieve miraculous results.

Adding heat to anything gives it energy to move around more fluidly, and gases react more enthusiastically than any other Ingredient. Expanding gases inflate marshmallows, doughnuts, and bread. They burst tomato and pepper skins, sausage casings, broad beans, and popcorn kernels. Any time we heat food to expand gas, we need to strike a balance between the gas and the other Ingredients. Baking bread with a good rise requires proper timing. We want gas to inflate the loaf before the outer crust gets too dry to let it expand. Pommes soufflées, puffed potato chips, require potato slices thick enough to catch escaping gas yet thin enough to be able to puff with the force. Adding alcohol to frying batter creates ethereal, light textures because the liquid alcohol turns quickly into gas once it's in the fryer.

Pressure also plays a role. The most famous example is industrial white bread, which is made in a vacuum oven. These ovens were designed to leverage maximum expansion with less heat for a

> Gases expand and contract more aggressively than any other substance in the kitchen.

soft, moist loaf with very little crust. By using a vacuum to expand tiny air bubbles trapped in the dough, this technique also eliminated the need for yeast to leaven the bread. It was a shrewd strategy for speeding production and lowering cost, but it ultimately resulted in a product lacking the taste and aroma that yeasts contribute as by-products of the fermentation process. Pop Rocks are another example, made by infusing molten sugar with carbon dioxide and placing it in a vacuum chamber. As the gas expands, the sugar hardens, creating mini sugar bombs full of compressed carbon dioxide.

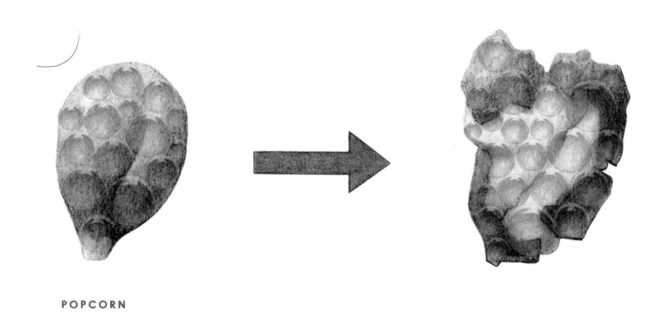

POPCORN

GASES / EXPANSION + CONTRACTION

When food is heated, gas bubbles expand, often growing several hundred times in size, which can cause it to leaven, puff, and explode.

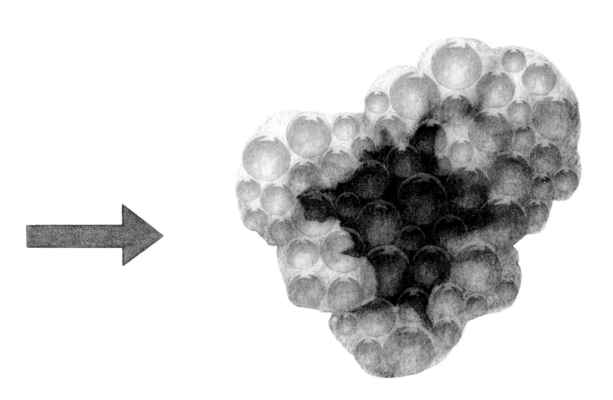

PORK SKIN CHICHARRÓN
Puffing happens when skin is cooked long enough that the softened protein scaffolding allows for expansion during frying.

GRILLED EGGPLANT
Skins on leathery vegetables, fruits, and meats burst during high-heat cooking because of steam expanding from within.

GASES / EXPANSION + CONTRACTION

OYSTER CRACKERS

Oyster crackers, pita bread, pommes soufflées, and other puffed foods have to be thick enough that carbs and proteins can trap gas, but not so thick that the trapped gas can't push out enough to expand.

TEMPURA BROCCOLI

The shape of tempura-fried foods comes from the tumultuous environment of frying, as water escapes.

MACARONS

Heat makes the water in puffed confections expand. Once the heat and expanding pressure is gone, carb and protein gels take over to preserve the shape.

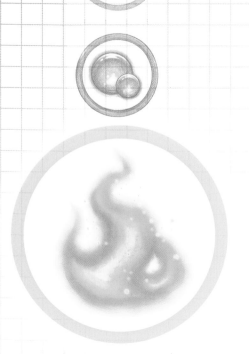

digital

HEA

T

Heat is energy, not a physical molecule like the other Ingredients. If water is the theater in which the other Ingredients perform, heat is the conductor and director of the show. With a combination of temperature and time, heat sets the tempo of every process in this book. It may seem like an abstract concept, but heat can really only affect food in two ways:

- It makes things **MOVE FASTER**.
- It makes things **VIBRATE FASTER**.

MOVE

Heat gives Ingredients a boost as they slide, float, zip, and roll around. With heat, Ingredients move faster, affecting everything from texture to flavor release.

Heat makes solid food softer. Most of the Ingredients are still locked into place, but they bend and shift more easily when we nudge them. Nut butters, cream cheese, and chocolate become more spreadable. Thawed ice cream becomes more scoopable, and thawed meat becomes more sliceable. Crispy foods become rubbery as Ingredients wriggle around rather than shattering like glass when we bite them. Tempered cheese feels creamier in the mouth, and even grainy, overcooked quiche feels smoother when it's served warm.

Controlling heat gives us the ability to turn a static concept like "solid" into a dynamic spectrum. Moving solid food to the colder side of the spectrum allows us to hold food in place while we cut ultrathin slices and grind big pieces into fine powders and tiny granules. Less heat means more crispy and more crunchy, and more chewing required to break stuff apart. The other, hotter side of the spectrum allows us to spread, smear, bend, and pull things more easily. Heat softens edges; favors natural, rounded shapes; and makes food seem light and ethereal as it moves more easily out of the way of our tongue.

Heat makes liquid food thinner. It gives Ingredients in a crowded liquid mob the energy to shove past one another more quickly. Sticky, tacky things like molasses, honey, and caramel become pourable syrups. Greasy fats turn into slippery, smooth butter sauces and bacon drippings. Heat gives water the ability to duck and dive around carb and protein roadblocks, thinning tomato sauces, cheese fondue, and starchy mashed potatoes.

Heat creates a spectrum of liquid texture for us to play with as well. On the cold side of the spectrum, things become lumpy, gelatinous, sticky, gummy, and greasy. Less heat means that sauces seize on a cold plate, syrup sticks in the bottle, and emulsions achieve maximum creaminess. On the hot side of the spectrum, things become runny. Heat favors free-flowing shapes and textures that are smoother, cleaner, less tacky, and less gummy.

Gases expand and travel faster. Gases are so much empty space already that we don't really notice a spectrum in their texture, but we do notice differences in the effects they have on solids and liquids. Hotter gas makes things expand and puff more dramatically. Puffed cereal, chips, popcorn, croissants, pizza, and *gougères* owe their textures to expanded gas.

With all of these physical changes, heat brings taste and aroma to our noses and tongues quicker. Solids and liquids hold on to their dissolved sweet, sour, bitter, salty, and umami components less tightly. This makes it easier for them to slip down onto our taste buds and make their existence known. Heat also helps gaseous aromas travel to our noses quicker. This means that hot food loses aroma more quickly. An important caveat: Smells are *not* smart bombs. They don't care where your nose is, and they will float off into space if you aren't around. This is part of the idea behind pressure-cooking: Seal everything into a container when food is hottest, and open the container only when everything cools down and aromas are less likely to fly away. Our obsession with cold brewing coffee and other drinks came from this realization—by never exposing food to heat, we have the chance to keep aromas from flying away and create a more aromatic product. On the other hand, heat helps aromas come out of the coffee beans, tea leaves, or whatever we're steeping in the first place. Is it better to have smelled and lost? Or never to have smelled at all? Everything is about balance.

Getting mixtures of Ingredients to stay evenly mixed can depend heavily on heat. Sugars, minerals, and other water-loving Ingredients dissolve better with heat, since water can move around them more effectively with the extra boost of energy. Fragile mixtures, like emulsions and foams, suffer from heat; in these mixtures, lipids and gases are held in contact with water against their will, and heat gives them the energy to move more quickly in search of an

> Getting mixtures of Ingredients to stay evenly mixed can depend heavily on heat.

We could
heat and cool
food forever, taking it
through the same cycle
for all eternity without
change, were it not for
vibraion..

escape route. This is why hot foams and emulsions are some of the most impressive nature-defying things we do in the kitchen.

These trends apply to every food you will ever eat. "Heat makes stuff move faster" is one of the most fundamental rules in the universe. The only other thing you have to consider when managing heat in food is that, in addition to moving faster, heat makes Ingredients *vibrate* faster. We could heat and cool food forever, taking it through the same cycle for all eternity without change, were it not for vibration.

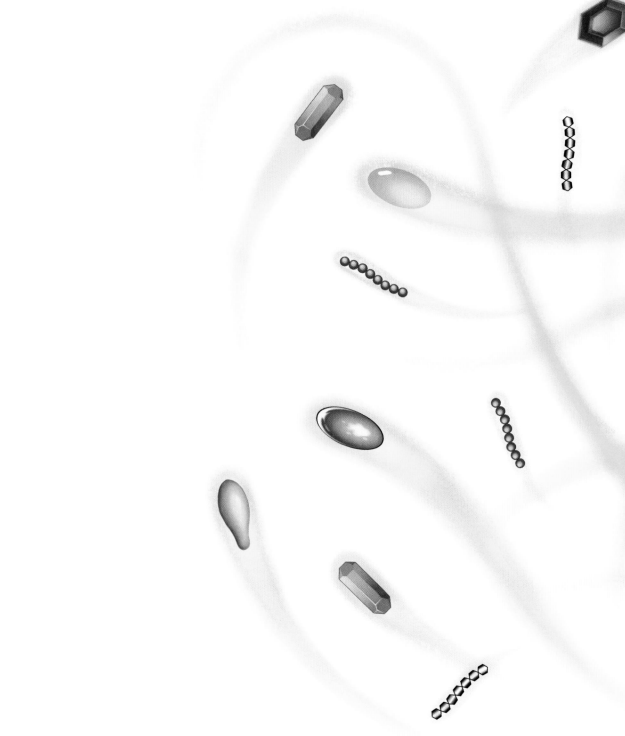

HEAT / MOVE

Heat causes the other Ingredients to move around faster, changing everything
from the texture of food to the way that aromas reach our noses.

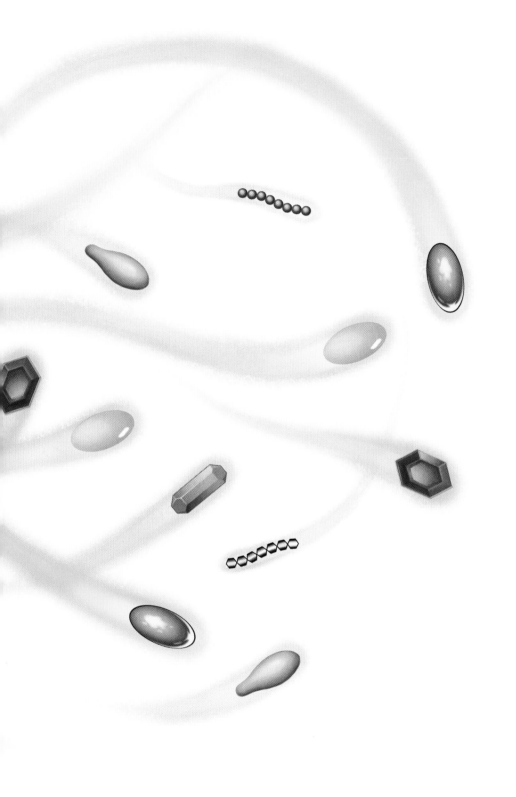

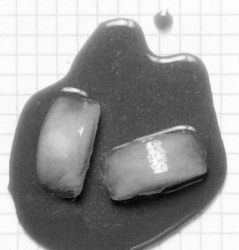

CHEESE PIZZA

Bubbles on pizza require heat to turn water into steam and loosen the cheese enough to bubble.

COLD-BREW COFFEE

Varying the heat during steeping of coffee beans changes the mixture of molecules that wander out into the water, thus changing the flavor.

HEAT / MOVE

CHOCOLATE HAZELNUT SPREAD ON TOAST

The ability to spread anything means balancing enough heat to make it spreadable while not adding so much that it loses its shape completely and runs everywhere.

ICE CREAM

In restaurants, the best pastry chefs are good at managing containers of ice cream so that nothing gets too cold to scoop nor too melted to hold a nice shape.

TOMATO POWDER

Sun-dried tomatoes are leathery, turning into a paste when blended, but they become brittle when chilled, so that they can be ground into a powder.

VIBRATE

Heat makes Ingredients vibrate. Vibrating Ingredients filled with energy are more likely to change. They shift shapes, they stick together, they fall apart, and sometimes they just plain explode.

Some of the good stuff in food is fragile, and heat can snuff those things out of existence. The delicate aromas of fresh basil, cantaloupe, jasmine, and strawberries die at the mere mention of heat. Colorful pigments can break apart, which is why "rainbow stew" is not a thing. Texture changes too, as Ingredients like proteins unfold and lash wildly about, globbing on to one another and forming new structures. In the previous section, we talked about how heat makes the other Ingredients move faster, which normally makes food flow easier, but we sometimes see food *thicken* with heat, as these new structures create bigger roadblocks. Taste is often sturdier than aroma or color, so it's often the only thing that remains unchanged after a lot of heating. Canned soups are a prime example of this, since many of them are loaded with salt, sugar, and umami to compensate for the loss of aroma to the intense heat of the canning process. With enough heat, however, even taste

> Vibrating Ingredients filled with energy are more likely to change.

can change. Sweet things can become sour and bitter, bitter things can break apart to become savory, and so on.

Out of the ashes, new deliciousness is sometimes born. Heat lights the fuse to make sugars and proteins explode, causing Maillard browning and caramelization. The dark, deep flavors of tomato paste and strawberry jam are a pleasant counterpoint to fresh tomatoes and strawberries.

Heat also controls enzymes. Enzymes, and the microbes that create them, work faster and faster with heat, until the point where they fall apart. Heat gives them more energy to work faster and harder until they start to overheat. There is a tiny window of heat exposure, a bliss point, when they reach the top of the hill before falling down the other side to their doom. That is the best way to describe heat: Undershoot and wallow in inefficiency; overshoot and get burned.

HEAT / VIBRATE

Heat causes the other Ingredients to vibrate, and when they vibrate
too hard, they can start to break apart and transform.

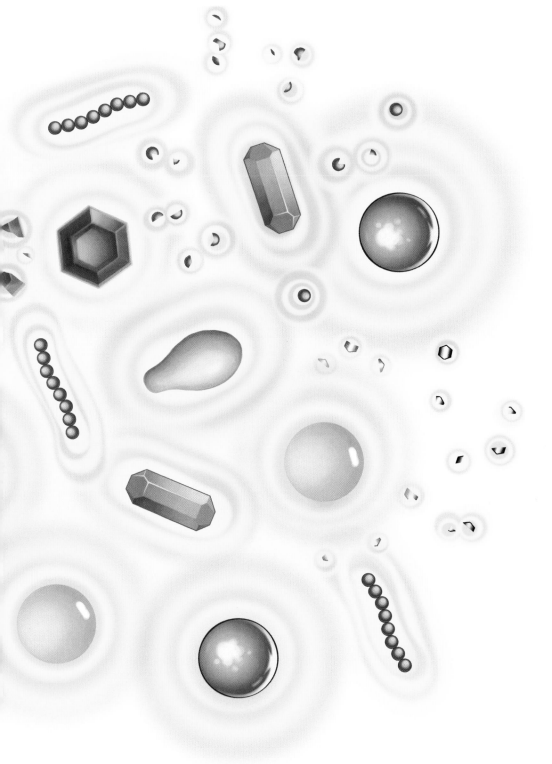

CRÈME BRÛLÉE
The high heat of a blowtorch allows the surface of crème brûlée to brown and crispify quickly without overcooking the custard underneath.

TOMATOES

CILANTRO
Green, floral, and fruity aromas are among the first to die when food is heated. Tomatoes take on a mellower, deeper flavor, but cilantro has nothing left to give.

HEAT / VIBRATE

CANNED SOUP

The intense heat of industrial canners discombobulates the proteins inside microbes that help them live. Unfortunately, that same heat also causes taste and aroma to break down.

CARAMELIZED ONIONS

—

SCRAMBLED EGGS

Onions and eggs are two classic examples that show how cooking things at low heat for a long time can produce very different results from high heat for a short period of time. It's not just the total amount of heat that goes in, but how we budget it out over time that matters.

THE NEW PANTRY

Now that you've seen how food works, you're going to notice the strings that control food in your breakfast, at your favorite restaurant, in a bag of chips, on cooking shows, everywhere. The easiest way to start pulling those strings yourself is to organize your mental pantry in a new way. Start thinking about the ingredients that you cook with in terms of which Ingredients within are most important for a particular dish. If you're trying to make glazed short ribs, something is going to have to make that sauce thick enough to stick to the meat, so where are the carbs and/or proteins going to come from? If you want a biscuit to turn golden brown in the oven, you'll need to brush it with something containing sugars and proteins, but what are your options? This section contains a list of typical sources of each Ingredient to get you started. This isn't an exhaustive list but more of a lineup of usual suspects. Start here, and as you get used to this new way of seeing, you'll discover more examples to add to your new pantry.

The list below includes foods that contain "a lot" or "some" of each Ingredient. Food comes from living things that rely on all eight Ingredients to survive, so everything you will ever eat probably contains at least a couple of molecules of all seven physical Ingredients and some amount of heat. This list focuses on sources that have enough of these Ingredients to make a noticeable difference in how a dish will turn out.

WATER helps you make liquids thinner, make things dissolve, make emulsions more stable by giving droplets more room to breathe, leverage pH to make things acidic or basic, make food firmer by freezing and crystallizing, and provide expansion by turning into steam.

A LOT: *vegetables, fruit, meat, seafood, eggs, herbs, fungi, vinegar, wine, beer, juices, soda, milk, condiments (soy sauce, fish sauce, ketchup, etc.), stock/broth, fresh cheese*

SOME: *aged cheese, syrups (honey, maple, etc.), thick emulsions (butter, mayonnaise, cream, etc.), dried meat, dried fruit, dried vegetables*

 SUGAR helps you make things sweet, make liquids slightly thicker or stickier, form crispy glasses when water is removed, give microbes food for fermentation, make something brown when heated, keep water from making coarse ice crystals in frozen foods, and keep water from evaporating out of cooked food.

A LOT: *granulated sugars, syrups (honey, maple, agave, etc.), jams and preserves, soda, sweet wine, fruits, beets, sweet potatoes, sweet corn*

SOME: *other vegetables, dairy, meat, fungi, seafood, grains, legumes, beer, wine, vinegar, bread, eggs, nuts, coffee*

CARBS help you make liquids a lot thicker, form crispy glasses when water is removed, bind mixtures together, create gels, stabilize emulsions and foams, and act as a reservoir for sugars that can be broken down later.

A LOT: *grains, legumes, nuts, seeds, root vegetables, bread, fruits, seaweed*

SOME: *leafy vegetables, herbs, spices, fungi*

LIPIDS help you store aroma and color in food, provide droplets to make emulsions, sometimes act as emulsifiers to preserve emulsions, help with heat transfer, create creamy textures when liquid and firm textures when solid, generate aromas by breaking into flavorful bits, and keep water-loving things like proteins and carbs separated to create tender batters and doughs.

A LOT: *oils, fats, meats, dairy, nuts, avocado, thick emulsions, chocolate, seafood*

SOME: *spices, fruits, vegetables, herbs, flour, seeds, grains, legumes*

PROTEINS help you make liquids a lot thicker, form crispy glasses when water is removed, stabilize and emulsify emulsions and foams, create a gel or bind something together, cause things to turn golden brown, and bind tastes and aromas.

A LOT: *meat, seafood, legumes, dairy, nuts, eggs, seeds, grains, wheat flour*

SOME: *fruits, vegetables, spices, herbs, fungi*

MINERALS help you make things salty, keep water away from microbes, prevent water from making coarse ice crystals in frozen foods or evaporating out of cooked foods, help large Ingredients like carbs and proteins link up to form gels, and give food color.

A LOT: *salts, aged cheese, celery, shellfish, fermented soy products, tofu, pickles, capers*

SOME: *vegetables, fruit, meat, chocolate, dairy*

 GASES help you thicken liquids with bubbles, make food puff with expansion, and make food react.

A LOT: *fizzy beverages, dry ice, fermented foods, yeast*

SOME: *fruits, vegetables, spices*

 HEAT helps you make food more pliable, thin, and freely moving; make any process happen faster; and kill enzymes and microbes if used in large enough quantities. This book focuses on the fundamental roles that heat plays in food; there are plenty of other great books that have been written on specific techniques and equipment for getting heat into food.

Ingredient was designed to help you see the handful of simple patterns at play behind the vast universe of dishes that we cook and eat in kitchens all over the world. While these eight Ingredients are simpler than the thousands of ingredients they comprise, there are still a lot of concepts in this book, and it can feel like a lot to retain.

I was never good at history—it always felt like there were too many random facts with nothing solid to cling to. If I had gotten to know Vasco da Gama as a person with hopes and dreams rather than just a Portuguese name, a date from the 1500s, and some bullet-point factoids, I would probably have an easier time remembering why he was important.

The easiest way to make this information stick in your mind is to focus on each Ingredient as a personality. Each of the eight Ingredients has its own signature style, and they come together like an ensemble cast of actors to tell the story of any dish. Proteins are unstable, dynamic, and finicky. Any stress, including heat, pH change, adding minerals and sugars, and physical abuses such as whipping, grinding, and mixing, causes them to lose their composure and change shape. Lipids are the other diva of the group. They stand up to heat better, but they have their own sensitivities like light and oxygen, which cause them to fall apart. Sugars are sturdier and more predictable. When they do change, they cause food to brown. They create deep, rich backgrounds of flavor rather than the attention-grabbing textural impact of disturbed proteins or the aroma takeover of degraded lipids. Minerals are even more stable. They don't break down, and they can't fly away—they stay where we put them. Minerals are as solid as the rocks that they come from. Gases are the polar opposites of minerals. Delicate and ethereal, they drift and billow with the slightest change in conditions, constantly striving to slip away from us. Carbs are probably the most boring Ingredient. They provide structure and support without the dynamic instability of proteins. The most exciting thing they can do is break apart into sugar. Water is the stage, the theater in which all of these players perform. Any change in water will invariably affect the entire cast. Heat is the conductor/director—we never see it, but it drives the pace and energy of the performance.

Cooking is a complex and personal affair. We all have different styles, abilities, resources, interests, and traditions that influence the dishes we want to cook. Knowing the personalities of the eight Ingredients is the key to having more confidence to create the food that we want to eat. By clearing some of the guesswork and voodoo out of the way, we can spend more time enjoying our food rather than fretting about how it will turn out. Whether you want to unshackle your creativity to come up with new and exciting ideas or you just want to follow a recipe and have it taste as good as the picture looks, these Ingredients are what make it all happen. Whatever the story is, and however you want to tell it, these characters are there to act it out for you.

Happy cooking.

ACKNOWLEDGMENTS

Development and Logistics
Pilot R+D (Dana Peck, Dan Felder, Kyle Connaughton, and George Volkommer)

Concept Design
Kelsy Wood
Conor Wood
Corey Murphy
Ethan Hart
Peter Wynn

Culinary Director
Sean McGaughey

Editorial Support
Louise Bouzari
Elizabeth Alderfer

Expert Review
Guy Crosby
Jeff Potter
Christopher Loss

Photography
Jason Jaacks

Illustrations
Jeff Delierre

In addition to all of the people who had a direct hand in making this book, thanks to all of the amazing chefs whose collaboration inspired the evolution of this way of thinking about food: Thomas Keller, Corey Lee, Brandon Rogers, Cortney Burns, Nick Balla, Matthew Accarrino, Daniel Humm, Christopher Kostow, Stuart Brioza, Nicole Krasinski, Maxime Bilet, John Shields, Dave Beran, Devin Knell, Bryce Shuman, Eamon Rockey, David Nayfeld, Kevin Farley, Alex Hozven,

Wylie Dufresne, Paul Liebrandt, Gunnar Karl Gislason, Janine Weismann, Chad Robertson, Aaron Koseba, Tim Bender, Phil Tessier, and Francisco Migoya.

Thanks to Harold McGee for my undergraduate education (I got more out of reading *On Food and Cooking* cover to cover as a biochemistry major than I would have gotten out of four years of food science.)

Thanks to Francis Lam for convincing me that this book was a good idea and to Jonah Straus for convincing everyone else.

Thanks to Diane Barrett, Tyler Simons, and UC Davis Food Science for teaching me how to ask questions and letting me do a PhD on mashed potatoes.

Thanks to my friends for being designers, musicians, programmers, artists, actors, scientists, and cooks, and showing me that thinking in three ways is better than one.

Thanks to my food-obsessed Texranian family, and a special high five goes to my sister, Meri, for searing food into all of my favorite memories.